# Cities and Consumption

*Cities and Consumption* investigates the mutual and dynamic relationship between urban development and consumption. Made accessible for students are debates that posit consumption as being at the intersection of different spheres of everyday life – between the public and the private, the political and the personal, the individual and the social. Consumption is thus understood as having multiple political, economic, social and cultural roles, and it is in the morphology of cities that its expression is most explicit.

*Cities and Consumption*:

- provides a critical review of the ways in which urban consumption has been conceptualised;
- describes the relationship between consumption, the development of the modern and postmodern city, and associated archetypal spaces, places, identities, lifestyles and forms of sociability;
- looks at both ordinary and spectacular consumption;
- explores the relationship between consumption and class, gender, ethnicity and sexuality;
- discusses how cities are consumed through representations in popular cultural forms and official discourses, and how cities themselves are consumed;
- shows how consumption is central to the ability of cities to be successful in a contemporary urban hierarchy characterised by intense competitiveness.

Using case studies and illustrations from North America, Europe and Asia, *Cities and Consumption* is an essential student text. It clearly presents key ideas, perspectives and ways of approaching the complex relationship between urbanity and consumption.

**Mark Jayne** is a Lecturer in Human Geography at the University of Manchester.

# Routledge critical introductions to urbanism and the city

Edited by Malcolm Miles, University of Plymouth, UK
and John Rennie Short, University of Maryland, USA

International Advisory Board:

| | |
|---|---|
| Franco Bianchini | Jane Rendell |
| Kim Dovey | Saskia Sassen |
| Stephen Graham | David Sibley |
| Tim Hall | Erik Swyngedouw |
| Phil Hubbard | Elizabeth Wilson |
| Peter Marcuse | |

The series is designed to allow undergraduate readers to make sense of, and find a critical way into, urbanism. It will:

- cover a broad range of themes
- introduce key ideas and sources
- allow the author to articulate her/his own position.
- introduce complex arguments clearly and accessibly
- bridge disciplines, and theory and practice
- be affordable and well designed

The series covers social, political, economic, cultural and spatial concerns. It will appeal to students in architecture, cultural studies, geography, popular culture, sociology, urban studies, urban planning. It will be trans-disciplinary. Firmly situated in the present, it also introduces material from the cities of modernity and postmodernity.

**Published:**
Cities and Consumption – Mark Jayne

**Forthcoming:**
Cities and Cultures – Malcolm Miles
Cities and Cinemas – Barbara Mennel
Cities and Nature – Lisa Benton-Short and John Rennie Short
Cities, Politics and Power – Simon Parker
Digital Cities – Chris Benner
Cities and Economies – Yeong-Hyun Kim and John Rennie Short
Urban Erotics – David Bell and John Binnie
Children, Youth and the City – Kathrin Hörshelmann and Lorraine van Blerk

# Cities and Consumption

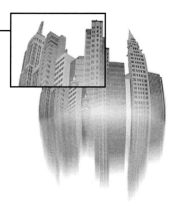

Mark Jayne

 Routledge
Taylor & Francis Group

LONDON AND NEW YORK

First published 2006
by Routledge
2 Park Square, Milton Park, Abingdon, Oxon OX14 4RN

Simultaneously published in the USA and Canada
by Routledge
270 Madison Ave, New York, NY 10016

*Routledge is an imprint of the Taylor & Francis Group*

© 2006 Mark Jayne

Typeset in Times New Roman by
Keystroke, Jacaranda Lodge, Wolverhampton
Printed and bound in Great Britain by
TJ International Ltd, Padstow, Cornwall

*British Library Cataloguing in Publication Data*
A catalogue record for this book is available from the British Library

*Library of Congress Cataloging in Publication Data*
Jayne, Mark, 1970–
Cities and consumption / Mark Jayne.
p. cm. – (Routledge critical introductions to urbanism and the city)
Includes bibliographical references and index.
ISBN 0–415–32733–4 (hardcover : alk. paper) – ISBN 0–415–32734–2
(papercover : alk. paper). 1. Urbanization. 2. Cities and towns.
3. Consumption (Economics) I. Title. II. Series.
HT361.J27 2005
307.76–dc22                                                      2005003655

ISBN10: 0–415–32733–4       ISBN13: 9–78–0–415–32733–6 (hbk)
ISBN10: 0–415–32734–2       ISBN13: 9–78–0–415–32734–3 (pbk)

# Contents

# Figures

# Tables

# Case studies

Case Studies

# Acknowledgements

There are many people to whom I am very grateful for their help and support during the writing of this book. I would like to thank my family and friends, particularly my Mum and Dad who have supported and encouraged me over the years. I would like to thank staff and students at Staffordshire University and at the University of Birmingham, in particular Wun Chan, Tim Edensor, Ann Gray, Jon Hill, Ruth Holliday, Brian Jacobs, Derek Longhurst, Maggie O'Neill, Tracy Potts, Tony Spybey, Frank Webster and Gemma Boon. Special thanks go to David Bell for his constant help, support, advice, guidance and patience. I would also like to thank other colleagues, Jon Binnie, Tim Hall, Sarah Holloway, Ellie Jupp, Justin O'Connor and Gill Valentine who have been helpful and supportive in a number of different ways. I would like to thank the series editors, Malcolm Miles and John Rennie Short, for giving me the chance to contribute to the series, as well as Andrew Mould, Anna Somerville and Zoe Kruze at Routledge, who have reacted to my missing numerous deadlines, and the various elaborate excuses I provided, with great patience and encouragement. I would also like to thank the anonymous referees for their insightful suggestions and comments that have helped to enhance this book. Special thanks to Ruth, David and Daisy.

The author and publisher thank the following for granting permission to reproduce the following material in this work:

Selfridges and Scott King for Figure 1.1 'Shop Until you Drop', summer 2003.

Sage Publications Ltd for Figure 1.2 'The Circuit of Culture' from Du Gay, P. (1997) *Production of Culture/Cultures of Production*, p. iv.

American Academy of Political Science for Figure 2.1 'Burgess's concentric zone model'.

American Academy of Political Science for Figure 2.2 'Hoyt's sector model'.

FLC/ADAGP, Paris and DACS, London for Figure 2.3 'Le Corbusier's vision of the contemporary city'.

Simmons Aerofilms Ltd for Figure 2.4 'A suburban estate, Newcastle upon Tyne, UK'.

Malcolm Miles for Figure 3.1 'Manhattan seen from Battery Park' from Miles, M. (1997) *Art Space and the City*, London: Routledge, p. 22.

Taylor & Francis Books, Inc. for Figure 3.3 'The Luxor, Las Vegas' from Rothman, H. (2002) *Neon Metropolis*, p. 196.

Anne Friedberg for Figure 3.4 'Europa Boulevard' in the West Edmonton Mall, 1992'.

Ken Mackay at Hurford Salvi Carr for Figure 3.5 'Loft living' from *The Independent* (2004).

Paul Chatterton for Figure 5.1 'Regulation aimed at tackling drunkenness in public space' from Chatterton, P. and Hollands, R. (2003) *Urban Nightscapes: Youth Culture, Pleasure Spaces and Corporate Power*, London: Routledge, p. 63.

Stephen Pope for Figure 5.2 'A black music specialist record shop', 1999.

Iain Borden for Figure 5.3 'Skateboarding in the city' and for Figure 5.4 'Benches designed to deter skateboarding' from Borden, I. (2001) *Skateboarding and the City*, London: Berg, p. 225 and p. 185.

Tim Edensor for Figure 6.2 'Cooking on an Indian street' and Figure 6.3 'A barber's on an Indian street'.

The Black Country Living Museum for Figure 6.4 'Heritage in the city'.

Eureka Entertainment for Figure 6.5 'An image from the film *Metropolis*'.

Tim Hall for Figure 3.2 'The post-industrial "global" metropolis', Figure 7.1 'The International Convention Centre, Birmingham, UK' and Figure 7.2 *Forward*.

T. C. Chang for Figure 7.4 'Singapore's central cultural regeneration' from the *International Journal of Urban and Regional Studies* (2000).

Charlie Scheips for Figure 7.5a and 7.5b Chicago Millennium Park.

Every effort has been made to trace copyright holders: any omissions brought to the publisher's attention will be remedied in future editions.

# 1 Introducing cities and consumption

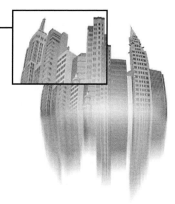

**Learning objectives**

- **To understand the relationship between consumption and urban change**
- **To describe theoretical conceptions of consumption**
- **To think about definitions of the city**

*Cities and Consumption* investigates the mutual and dynamic relation between urban development and consumption. Over the past twenty years, literature relating to the study of cities and consumption has produced a significant contribution to our understanding of the contemporary world. Linked to conceptualisations of increasing global interdependence, the study of cities and consumption helps to explain seemingly globally ubiquitous change – relating to political life and governance; economic restructuring and changes in employment; the type of spaces and places that have developed in our cities – as well as shedding light on changes in the everyday realm of our social and cultural lives, including where and how we spend our leisure time; where and what we eat; where we go on holiday; where and when we do our shopping; what we wear; how we decorate our homes, and so on. In these terms, then, it is through the study of cities and consumption that a whole range of seemingly diverse but interconnected elements can be brought together. This allows us to conceptualise and describe broad urban changes and also to ground them in the knowledge that we all actively constitute and are participants in city life in unique ways.

Consumption stands at the intersection of different spheres of everyday life – between the public and the private, the political and the personal, the social and the individual. Consumption is understood to be a means and motor of economic and social change; an active constituent in the construction of space and place; and as playing a vital role in constructing our identities and lifestyles (Miles

1998a). In these terms, consumption has multiple political, economic, social and cultural roles, and this book will show that it is in the morphology of cities that its expression is most explicit – cities, and places and spaces within them, are the sites in which consumerism has been spectacularly mediated.

For instance, the rapid growth of cities throughout the world, in terms of both population size and land use (often characterised as 'urban sprawl') has led to an often dramatic physical restructuring, underpinned by changes in the way cities are governed, and new social and cultural relations. New spaces and places – such as revitalised city centres characterised by gleaming high-rise office buildings, flagship buildings, waterfront developments, tourist, ethnic and cultural quarters, and affluent suburbs – are representative of urban success stories. In contrast, there has been an increasing social and spatial polarisation between the 'haves' (who occupy such new dramatic urban spaces and places) and the 'have-nots', often found in economically poor areas or ghettos where segregation is based predominantly along the lines of class and ethnicity.

There is intense global competition amongst cities for investment, jobs, tourists and cultural facilities and infrastructure; in this competition there have been places and people who have been winners, and others who have not benefited. The way that urban theorists have conceptualised change in different cities and in terms of different groups of people throughout the world has brought a high profile to urban studies. This has ensured that urban theory has become central to the engagement of social scientists in contributing to wider debates about the changing world.

While urban studies as a whole has been invigorated by such real-world change, the past two decades has seen studies of consumption move from being an oft-ignored aspect of our lives to having a central berth in the minds of theorists. This heightened interest has again been bound up in the complex set of changes in political, economic, social, cultural and spatial practices and processes that are constituted by, and impact on, everyday experience and preoccupations. For example, for most of us our understanding of what 'consumption' is, is a long way from its historical meaning of to exhaust, destroy, to use up, a definition with such wasteful negative connotations that 'consumption' became a vernacular word to describe pulmonary tuberculosis (TB) – an often terminal lung disease prevalent in industrialised nations up until the 1950s.

Today consumption is more likely to be considered as a pleasurable pursuit, often involving imagination and the realisation of desires and ambitions – it's the stuff of shopping malls, 'retail therapy', Christmas credit-card binges, designer boutiques, trendy restaurants and bars and foreign holidays. Conversely, consumption can be synonymous with exclusion – when we can't afford, or get access to, the things we want – and for many theorists consumption is as much about differential access to resources, as much about thrift shops and low-cost super-markets, as it is about high-profile spectacular spaces and places in our cities. One thing is clear, however: consumption constitutes the 'stuff' that surrounds us all – advertising, television, entertainment, shopping – and it is consumption that underpins the images, sounds, smells and sights of the contemporary world. In these terms, then, for many theorists the ever-increasing and differentiated consumption opportunities have become the defining characteristic of twenty-first-century life.

This book shows that the relationship between consumption and *urban life* has been a central component of the development of a consumer society. Under-pinning this close relationship over the past thirty years has been the re-orientation of city life away from the logic and organisation of the production of goods and services towards the position where consumption is the prime organisational feature. While this process of political, economic, social, cultural and spatial restructuring is not specifically an urban phenomenon, it is in cities that such change has taken place in its most concentrated form. Cities are important nodes where collective and individual consumption takes place on a massive scale. Indeed, for cities to compete in an urban hierarchy characterised by intense competition to secure investment, jobs and tourists, urban authorities have had to ensure that they create economically and symbolically important new urban consumer spaces. However, the relationship between consumption and cities is not solely measured in such symbolic and economic terms. Cities themselves are consumers of land, energy, water, and the raw materials that make up and sustain the physical infrastructure of urban life. Cities are consuming spaces as well as consumer spaces.

It is in these terms that we can see that the contemporary city is defined by and through consumption. The contemporary city is the product of the post-industrial consumer economy created through a fundamental shift between production and consumption – where the balance of power between the producer and consumer that pervades daily life has changed in favour of the latter. As such, the con-temporary city is currently defined by a belief in consumption, and the current practices and future intentions of urban imagineers, local authorities and private-sector producers (financial speculators, architects, urban planners and other council officials) disseminate a clear message that it is consumption that is set to shape the future of our cities. Consumption has become so ingrained in both political and institutional visions, planning and policy regimes and our everyday lives that consumption is not simply about goods and services but is central to our vision of the city, the very idea of the city. It is in these terms that understanding of the contemporary city and indeed visions for the future of the city are bound up with ideologies of consumer sovereignty and choice.

However, while consumption is omnipresent and plays an increasingly central role in people's lives it is important to consider what consumption offers to cities and to the everyday lives of its inhabitants. It is clear that, for many cities, consumption-led economic development strategies offer possibilities for competitive economic advantage, or simply represent an attempt to halt economic decline – but at what consequence? The relationship between cities and consumption is contradictory: for example, while some people do better than others in the consumer society, everyone is a part of it – consumption benefits some consumers but at the expense of others; consumption divides and excludes as much as it provides opportunities. While consumption can be progressive, it creates as many problems as it seems to solve. As such, a key question posited in this book is to assess the extent to which consumption is a positive force in the city. This question runs through each chapter in order to question how consumer society and consumption impact on the character and form of a diverse range of cities (and spaces and places within them) and the different social groups that constitute urban populations.

## What do we mean by consumption?

Before moving on to a more detailed introduction to how the specific relationship between cites and consumption has been conceptualised, it is important to pause in order to get to grips with what academics actually mean when they talk about 'consumption'. Most of us are aware that we are consumers of various things, that we have certain consumer rights and that our lives, the things we do, the places we go, and the places we live in are structured to a large degree around our consumption needs. Despite this useful common-sense appreciation of consumption, a satisfactory theoretical conception is more difficult to capture.

Case study 1.1 provides examples of some of the different ways in which consumption has been defined. However, it is only when such descriptions are combined that they provide a comprehensive insight into the multifaceted definition that academics are using when they talk about consumption.

### Case study 1.1 **Defining consumption**

Consumption has been described as being simultaneously about the 'selection, purchase, use, reuse and disposal of goods and services' (Campbell 1995: 104); as 'comprising a set of practices which permit people to express self identify, to mark attachment to social groups, to accumulate resources, to exhibit social distinctions, to ensure participation in social activities' (Warde 1997: 304); and as central to the ways in which we construct, experience, interpret and use spaces and places (Urry 1995).

As such, consumption is not just about goods that are manufactured and sold, but increasingly about ideas, services and knowledge – places, shopping, eating, fashion, leisure and recreation, sights and sounds can all be 'consumed'.

However, while such definitions provide a useful starting point to begin to delve deeper into the relationship between urban change and consumption, Stephen Miles (1998b) makes the important point that we must distinguish between, and make connections between, notions of 'consumption' and broader concepts of 'consumerism' and 'consumer culture'. These seemingly subtle distinctions are made clear in the idea that an old relationship between production and consumption, which was seen as the centre of modernist urban life, has now been reversed. Our identities and lifestyles are determined by our access to, and use of, an ever-increasing range of goods and services. The consequence of this is that social differentiation is argued to be based around our consumption practices rather than work and employment – in essence some theorists argue that how we spend our money is now more significant than how we make it.

Miles argues that consumption has taken on a 'magical' quality and that by becoming an important aspect in our everyday lives it is thus a thoroughly cultural phenomenon. In these terms, social theorists are interested in the kinds of influences, experiences and social relations that surround consumption as an everyday activity; hence the notion that consumerism has become not just an activity or pastime but a 'way of life'. This allows us to assert that there are 'hidden properties' of consumption (see Figure 1.1). In particular, such explanation exposes that there are ideological dimensions to the growth of a consumer society, and this allows us to show how 'the ubiquitous nature of consumption is reconstructed on a day-to-day basis' (Miles 1998b: 4).

This understanding of the daily reiteration of consumer ideology also allows us to critically reflect on the definition of consumption provided in Case study 1.1. As the definition suggests, consumption is about the purchase and use of goods and services; and studies have usefully considered how such practices help us to construct our identities, understand our place in the world and mark attachment to social groups, spaces and places. However, it is also acknowledged that goods and services are imbued with symbolic meanings that help us to communicate

Figure 1.1 *Selfridges' 'Shop Until you Drop' – a summer 2003 advertisement that playfully reveals the ideological dimensions of consumption. (Courtesy of Selfridges and Scott King)*

and interpret via consumption. As such, the symbolic content of commodities ensures that the way we ascribe meaning to them through consumption is to a certain extent 'up for grabs'.

For example, the way that both individual people and social groups consume goods and services relates to a matrix of identity positions (based around constructions of class, gender, ethnicity, sexuality, age, and so on). This ensures that commodities are 'unstable', to the extent that their symbolic content can be constructed, appropriated and (re)appropriated via symbolic meaning and hence 'consumed' in many different ways. It is in this sense that consumption can be a productive force, creating and (re)creating meaning and underpinning political, economic, social and cultural practices and values. The symbolic meaning bound up with commodities ensures that consumption enables a (re)negotiation and (re)working of the value, use and meaning of commodities and hence that consumption itself represents a productive and creative process.

It is also possible to extend this understanding to the idea that cities and the places and spaces within them are also consumed, particularly visually, but places can also be literally consumed (industry, history, buildings, literature and the environment over time can be depleted, devoured or exhausted by use). Thus, as city life has become increasingly oriented around consumption, the way in which we interpret, appropriate and (re)appropriate (consume) urban space is also bound up with our experience of everyday lives and concepts of value, use and meaning. Thus, while consumer capitalism unevenly structures our ability to engage in consumer society, the productive properties of consumption nonetheless ensure that the relationship between consumption and urban space and city life is also 'up for grabs'. Such a viewpoint points to the question as to whether contemporary urban consumer culture itself offers an opportunity to create fairer and more inclusive cities.

For example, it is argued that the contemporary city is characterised by conflict, social and spatial division, and conspicuous consumption. Large numbers of people are thus marginalised or often excluded from city spaces because they are economically not able to participate in consumer culture in the way, and to the extent, they would like. However, the idea that the city can offer unlimited opportunity for middle-class consumers and hence enforce social control via the dynamics of urban consumer culture is exaggerated. The reality of the consuming city is not unlimited opportunity for those who are economically more successful, nor is it that consumer culture inevitably generates entrenched social and spatial divisions. We are all members of urban consumer culture, and in reality there is a diverse mix of social groups that physically and symbolically occupy, produce meaning and create belonging in the spaces and places that constitute the commodified city.

Such an argument is not, however, based around a crude reproduction of a capitalist ideology of consumer sovereignty and choice. It is clear that global capitalism and its political, economic, social, spatial and cultural organisation ensure uneven opportunities for different social groups (and cities and spaces and places within them), and clearly consumption is a key element in structuring this inequality in daily life. Nevertheless, through an understanding of the transformative and productive capabilities of consumption, the daily production and reproduction of consumer culture offer new opportunities. These include, for example, new social and cultural solidarities; new, more inclusive practices and sociability in urban spaces; and potentially more socially just and environmentally friendly urban development.

For cities to work better for more of their inhabitants, what such a vision relies upon, of course, is the social will to embrace new social relations and forms of sociability. Central to this is the political impetus to instigate policies and planning regimes to provide services and develop an infrastructure that enables access to, and generates, a variety of consuming opportunities and hence potentially more inclusive urban spaces and places. A key theme throughout this book, then, is to show the ways in which consumption constitutes all our daily lives. This is illustrated not only through specific examples, but by identifying how urban consumption offers symbolic and practical opportunities to create more socially and environmentally sustainable cities. Chapters show not only the ways in which cities (and spaces and places within them) are engaged in creating competitive advantage, but how urban inhabitants are generating meaning and belonging for themselves through the productive force of consumption. A central endeavour of this book is to identify both the inequalities and disadvantages, and the more progressive political, economic, social, cultural and spatial practices and processes, bound up in the relationship between cities and consumption.

The remainder of this chapter will briefly outline the theoretical work that links consumption to urban change (as a primer to more detailed debate in later chapters). It concludes with an overview of the structure of this book and a brief description of each chapter.

## Theories of consumption and urban transformation

Up until the 1980s, consumption had been widely ignored by the social sciences. The recent renewed importance of consumption within social sciences reflects a view that 'people's actions and expectations as consumers have played an increasingly formative role in maintaining social life and that consumption could no longer be an afterthought to production processes' (Miles and Paddison

1998: 815). Such a view contrasts to that of classical sociological theory, which saw production and not consumption as having a central influence on the character of social life.

It is in the work of Karl Marx that this relationship was most famously espoused. Marx considered that it is production and people's relationship to the means of production that is the key determinant of social structures and relationships. Marx viewed the commodity as something to be sold and exchanged, and therefore having a role in determining social position. However, the capitalist mode of accumulating wealth is marked by the predominance of production for sale/exchange orientation as distinct from production for use, and it is our position within that system that marks our relationship with commodities.

For example, the capitalist mode of production features a market where commodities are bought and sold and where labour itself becomes a commodity, where wages are paid to workers dependent on either the time spent at work, or the amount of output produced. In this form of economy it is the capitalist who controls both the production process and labour. It was through the development of factories that a system proliferated whereby workers commonly lived near their place of work, often next door to their fellow workers, and together were part of a division of labour where individuals undertook specific and repetitive work. Rather than being actively involved with all of the production process, workers were only responsible for one small element. Marx argued that this produced a workforce 'alienated' from the commodities they produced. In this way, the capitalist retained authority and power over the workers, the environment of production and the choice of technique. For Marx, then, money becomes central to social formation, and central to the capitalist mode of production is competition. Competition forces and motivates the capitalist to secure and then to maintain their share of the market. Assets create wealth for the owner and embody and assert a relationship between those who make money and those who do not.

At the heart of this social relationship is the commodity – a product, produced by human labour, that satisfies human want. Under capitalism, then, a commodity takes on an exclusive exchange value (monetary value), produced by alienated labour. The object of labour (that is, the material artefact or product) thus has a crucial role in the construction of people's lives and, in turn, in their sense of well-being. In sum, the worker no longer has ownership of the commodity they produce, thus alienating them as a bit-part player, or is dehumanised. Marx describes this as a fetishism of the commodity, giving it mystical qualities, with a significance beyond its use value – to be treated with awe and reverence. However, the social relations of the capitalist production regime actively camouflage the exploitative social and labour relations that underlie the process as a whole. In these terms, what Marx understands of the commodity is contextualised purely by the production process. It is thus production rather than consumption on which he focused his argument.

Marx thus concluded that the process of 'commodification' underlies all aspects of social life. By workers earning wages or salaries, and being obliged to become embroiled in consumer culture, the wheels of capitalism are oiled by the workers. Commodities thus symbolise the power of capitalism, because the power structures that lay behind the commodity underpinned capitalism. Marx considered that it is production and people's relationship with the means of production that are key determinants of social structures and relationships.

This emphasis on consumption studies initially gained momentum in the early work of Manuel Castells (1977, 1978), David Harvey (1973) and later the ideas of Peter Saunders (1981). These theorists sought to depict the role of *collective consumption* (such as housing, education, health, sanitation, energy and other service provision provided by local governments to all in the city, in order to sustain the practical needs of industrial production, and to sustain workers in order to ensure economic growth – this will be returned to in subsequent chapters). For example, writing during the 1970s when a series of crises affected capitalism as a whole, Castells described a situation where the decline of traditional heavy industry and manufacturing, and the growth of new high-tech and service-oriented employment, were leading to massive loss of jobs. Castells argued that the effect of de-industrialisation and loss of wealth and revenue was leading to a reduction in city authorities' ability to pay via taxation for infrastructure, and that services would be hit as local government sought to balance budgets. Castells realised that it was collective consumption that would be first hit by this fiscal crisis. He argued that a popular labour movement would highlight workers' subordinate position in relation to the means of production and a revolutionary reconstruction would unfold.

Despite such Marxist analysis being grounded in laudable concerns for issues of social equity, in this conceptualisation consumption remained an add-on to the real task of theorising people's relationship to the means of production as a functional necessity for the reproduction of labour power and hence for the continued success of the capitalist mode of production. Nevertheless, this work did give important momentum to our understanding of processes of 'commodification', and led to a questioning of the assumption that the relationship between class and consumption was immaterial and that production was the primary means and motor of social organisation.

However, it would be a mistake to suggest that it was Marx and his followers who were the only early writers to acknowledge consumption as an important area of study. For example, writers such as Veblen (1899) and Simmel (1907) identified how consumption had developed as a marker of social prestige, and were engaged in discussing the geographies of the changing nature of cities. Later, theorists associated with the Frankfurt School (Adorno and Horkheimer 1973; Benjamin 1973) described how the growth of industrial manufacturing stimulated the development of urban population centres and exposed urban dwellers to an increasing plethora of goods. Writers associated with the Frankfurt School worked on Marx's concept of alienation and the way this emphasised that the culture industry integrated workers into the capitalist society. They argued that capitalism had produced a situation where the 'genuine' folk culture of the people had been destroyed by a highly commercialised and standardised mass culture. Culture had become something that was made and sold, just like any other industrial product, in order to make a profit. It was, according to this view, imposed on the masses by the culture industry, and turned people into passive consumers of material that did not meet their real needs. The Frankfurt School argued that mass culture was crucial to maintaining a capitalist society. In the capitalist mode of production, workers were duped into accepting boredom and exploitation at work through the idea that they could escape during their leisure hours into the pleasure of popular culture, shopping, watching films, and so on.

However, Bocock (1993) argues that coming out of the success of capitalist modes of accumulation – and fuelled by important events such as the widespread availability of credit cards – the 1950s saw the rise of a consumption culture that became more and more specialised. It is at this historical moment, it is argued, that fixed (structural) identity groups and social classes (such as those based on class, gender, ethnicity, sexuality and age) were beginning to be undermined, allowing individuals to 'freely' construct their identities and lifestyles through consumption choices (Featherstone 1991). Researchers thus began to consider the spatial patterning of consumption and how the social relations of consumption have become increasingly constructed through a tissue of particular city spaces/ places and associated lifestyles and identities.

An important concept in understanding how the importance of production and consumption to the individual's identity and sense of self shifted around this period is the figure of the 'affluent worker'. Goldthorpe et al. (1969) argue that as workers became better paid and more secure in their jobs, there was a shift in the way that work was viewed. While previously conceptions of status, self-worth and identity were solely based on the productive work they did on a day-to-day basis, there was now just as great an emphasis on the things that people could buy. This ensured a more instrumental view of work and a re-orientation away from a capitalist ethic of work-based identity to one based on consumption.

More recent writers such as Baudrillard (1975), Zukin (1982), Bourdieu (1984), Harvey (1989a) and Featherstone (1991) have also contributed to a body of literature concerning the history of the development of consumerism, consumption and the consumer society. It is through the work of these theorists that consumption gained prominence in high-profile academic debates which have sought to explain the rapid urban change that has unfolded in the past twenty-five years. This literature presented a new stage of consumerism and commodification and their symbolic and aesthetic relation to the emergence of a 'new middle class'. *Individualised consumption* and lifestyle patterns were thus part of a class distinction project (Bourdieu 1984). Alongside these developments, improved technology and manufacturing practices led to a decline in mass standardised industrial manufacturing and the growth in postFordist flexible specialisation, multi-skilled flexible workforces and small-batch production (see Case study 1.2).

These moves represent an attempt to overcome the over-accumulation weakness inherent in Fordist production – which resulted from too many mass-produced standardised goods being produced and left unsold. However, postFordism became more volatile as diversified market segments emerged to serve the interests of the consumer. In this new consumer world, consumption no longer appeared to be determined by the producer, and the producer (and ultimately the city) was increasingly subject to the demands and tastes of consumers. In this new regime of 'disorganised capitalism' (Lash and Urry 1994), consumption is seen as being less and less functional and more and more aestheticised. In sum, it was suggested that a shift in modern social life in terms of the balance between production and consumption had taken place. Hence, theorists have argued that consumption assumed a pro-active role in people's lives, becoming a focus of everyday existence regardless of their relationship to the means of production (Hebdige 1979; Featherstone 1991).

As such, consumption had become a focus for theoretical debates seeking to explain the cultural transformations which occurred in the light of the changing organisation of global capital accumulation and discourses of postmodern life. In effect, consumer choice has become the foundation for a new concept of freedom in contemporary society, in which individuals are socially constituted in terms of consumer ideologies and the ways in which they differentially construct meanings from such discourses (Leach 1984). Importantly, then, consumption can realistically be described as a bridge that links the individual to the urban environment. It is through consumption that people's activities and everyday lives and the physical organisation of the city are bound together (as well as constituting political, economic and social practices and processes). As such, urban studies is seen to have an important role in advancing concepts of the social significance of consumption.

## What do we mean by the city?

So far in this introductory chapter a number of different terms – post-industrial/ postmodern/postFordist – have been used to describe particular urban conditions. Such concepts seek to weave together complex physical, political, economic, social, cultural and spatial practices and processes in order to provide a comprehensive and generalisable account of urban change. This section unpacks these terms and reflects on their usefulness, but begins, however, with a discussion of what we mean when we talk about 'the city'.

Reflecting on the nature and definition of what a city actually *is,* Amin and Thrift (2002) suggest that cities have become extraordinarily intricate and that it is thus difficult to generalise about different cities. They argue that it is even difficult to agree on what counts as a city. For example, most cities now sprawl across many miles, incorporating settlements of varying sizes and compositions, and include a wide variety of spaces and places – derelict areas, parks and gardens, factories, shopping centres, car parks, warehouses, suburbs, dumps, houses, wooded areas, farmland, and so on.

Half the world's population now live in cities, and there are thirteen megacities with populations of more than 10 million (including Tokyo, New York, Mexico City, Shanghai, Mumbai, Los Angeles, Buenos Aires, Seoul, Beijing, Rio de Janeiro, Kolkata and Osaka). There are also numerous cities of national and regional significance, not to mention the many small cities that can be found throughout the world. Amin and Thrift (2002) thus argue that the city is every-where and in everything, and that we live in an urbanised world where a chain of metropolitan areas connects places through corridors of communication such as airports, railways, motorways, and virtually through information highways, in popular culture and in our imaginations. They also suggest that the footprints of the city are all over rural areas, in the guise of commuter villages, motorways, supermarkets, tourist destinations and an urban way of life. Amin and Thrift highlight that the divide between the city and the countryside has been perforated.

However, urban sprawl and the urbanisation of social life do not negate the idea that cities are distinct spatial formations. For example, Pile (1999) identifies three aspects that distinguish a city: the density of people, things, institutions and architectural forms; the heterogeneity of life that juxtaposes in close proximity; and the siting of various networks of communication that flow across and beyond the city. Similarly, Doreen Massey (1999) argues that it is the spatiality of the city, its density and juxtapositions of difference that produce a distinct urban pattern with diverse networks of interaction. As such, Amin and Thrift (2002) argue that the 'citiness' of cities seems to matter, although they suggest that it is now debatable how far spatial organisation remains a central feature of the sprawling global city. They conclude that the possibility of recognising cities as spatial formations gives us a legitimate object to study. Nevertheless, they conclude that no one city can tell us all about urban life and, despite similarities, there is extraordinary variety and complexity between different cities throughout the world. Despite this contention, there have been a number of theoretical attempts to generalise broad urban archetypal conditions. Conceptualisations of the city as modern and postmodern, Fordist and postFordist, industrial and post-industrial have been the most prominent ways in which theorists have sought to generalise urban change (see Case study 1.2).

## Case Study 1.2 **Defining the city**

The modern city is characterised by its homogeneous zoning of business and residential activity with a dominant commercial centre and a steady decline in land values away from the centre. Urban governance in the modern city is based on a managerial approach and involves the redistribution of resources for social purposes, and public provision of essential services. The economy is dominated by industrial mass production and based on economies of scale. The city features functional architecture and a mass production of styles, and is planned in its totality with space being shaped by social needs. There are strict class divisions, with a large degree of homogeneity within class groups.

The **postmodern city** is structurally chaotic, and hosts highly spectacular events in contrast to the presence of large areas of poverty. There is also an archipelago of distinct urban areas such as hi-tech corridors and post-suburban development. Urban governance is entrepreneurial, and resources are used to lure mobile international capital and investment. The public and private sectors work in partnership, and there is a market provision of services. The economy is service-based, with flexible production aimed at niche markets, based on economies of scope. The most important economic drivers are the globalised telecommunications and finance industries, and the city is oriented around consumption. Planning is undertaken in a piecemeal way, and urban design based on aesthetic rather than social ends. Architecture is eclectic and made up of a collage of styles that is often spectacular, playful, ironic and produced for specialist markets. Culture in the postmodern city is highly fragmented and characterised by lifestyles divisions. There is a high degree of social polarisation, and social groups are distinguished by their consumption practices.

The **industrial city** emerged as a result of manufacturing and production-focused capitalism. The nation-state holds close control of city power, and local authorities focus on development and maintenance of collective consumption infrastructure in order to service the needs of capital accumulation (and hence underpin class conflict). The physical organisation of the city is based around factory production, and the separation of the household from production. In order to increase levels of production and productivity, urbanisation is also predicated on improvements in provision of collective consumption and social welfare such as mass education, literacy, health and housing. It is industrialisation and not broader capitalist activity that determines the physical, political, economic and social trajectory of the city. The social structure of the city is dominated by extended family and friendship networks based on residential proximity.

The **post-industrial city** emerged during the latter part of the twentieth century and is characterised by a declining dependence on manufacturing and industry and a rise in the importance of service industries. There is also a particular emphasis on the role of knowledge in production, consumption and leisure. This includes an increased importance of higher education in order to stimulate a knowledge economy based on innovation. The social organisation of the post-industrial city is characterised by a dominance of knowledge-based professional, managerial and business occupational groups.

The **Fordist city** is organised around industrial production that is low-cost and standardised. The city is structured by residential and industrial zones in close proximity to each other, a central business district, and affluent residential areas, especially in suburban locations. The city is characterised by conflictual industrial and social relations. Workers' tasks are fragmented and repetitive, the labour force is occupationally divided, and management is centralised. The production process is dominated by Henry Ford's mass-manufacturing techniques and standardised products that generate mass consumption. Fordism is based on capital-intensive, large-scale plant, inflexible production, rigid bureaucracy, scientific management and semi-skilled labour. Political power and social structure are determined by this dominant mode of production and the social divisions of capitalists, managers and workers.

The **postFordist city** is characterised by a production landscape that has experienced the decline of the old manufacturing and 'smokestack' industries and the rise of new computer-based technology and more flexible decentralised labour processes and work. The city is dominated by the autonomy and needs of multinational corporations and the global processes of capital production that

continued

produce new international divisions of labour. Production in the postFordist city is underpinned by the economic possibilities of microchip technology, computers and the exchange of information and commodities. Economic activity is dominated by large numbers of small to medium-sized enterprises catering for segmented markets through flexible production of specialised goods and services. The power and spending of local government are restricted by central government, and non-elected bodies often determine economic development and planning policy. Associated social and economic transformations have led to a growth in the number of low-paid, insecure jobs and the promotion of consumption cultures pursuing the concept of individual taste and chosen lifestyles distinctiveness. These economic, social and cultural processes are mapped on to the city in terms of spectacular urban developments contrasted with areas of economic poverty.

It is clear, then, from the brief review of this trilogy of concepts in Case study 1.2 that theorists have sought to encapsulate particular constellations of practices and processes in order to describe and explain urban change. Each represents significant academic debate in its own right and, while often having much in common, each has its particular foci, and there is little consensus about which is the most comprehensive or satisfactory.

For example, theories of the post-industrial city have focused on issues of social structure, production, the service sector, occupation distribution, technology, job losses and a society fuelled by consumerism (see, for example, Bell 1976 and Douglas and Shaw 2001). Depictions of postFordism conditions meanwhile have addressed issues of identity, lifestyle and production (particularly a movement away from mass production to flexibility and 'disorganised capitalism' (see Lash and Urry 1987). Consumption, aestheticisation and class-derived aspects of space, politics, multinational companies, and individualism versus collectivity have also been pertinent concerns (Amin 1994; Lever 2001). Similarly, postmodern culture is described by Kumar (1995: 164) as 'the culture of a post-industrial society', and understanding it requires study of hybridity, social polarisation, fragmentation, hierarchy, planning and aestheticisation.

Despite the breadth, depth and variety of issues included in these different theories, all have been criticised for over-blowing the extent to which the conditions they depict have proliferated, and for ignoring their uneven and differentially constructed nature. Moreover, as Harvey (1989a: 189) suggests, while 'there has certainly been a sea-change in the surface appearance of capitalism since 1973 . . . the underlying logic of capital accumulation and its crisis tendencies remain the same'. However, whilst it is clear that none of these theoretical approaches is comprehensive, they all help us to understand in different (and often overlapping) ways what political, economic, social, cultural and spatial practices and processes are affecting cities. Measuring and judging the validity and usefulness of such theories is thus a relevant outcome of studying local contexts. As such, Miles (2001: 101) is correct to suggest that as a whole they 'may not be the best way of understanding the sorts of social changes that are going on around us. On the other hand [they] . . . may present the sort of metaphorical kick up the backside' that is needed to conceive the complexity of the urban condition. The right blend of theoretical and empirical study can bring a useful balance to studies of the city.

What is clear, however, is that while cities and urban life have developed over several centuries, no two cities are identical. While 'citiness' may be generalisable to a certain degree (and usefully so), individual cities have very different landscapes, economies, culture and societies. Cities are shaped by a diverse set of processes and these are dependent on factors that are unique to individual cities, such as size, nature of its economic activity, mix of social groups, and relationship to networks of cities and position in the urban hierarchy. As Hall (1998) summarises, urbanisation happens in different places at different times, with different speeds and to different degrees. Nevertheless, what concepts such as postmodern/postFordist/post-industrial urbanity offer is a useful generalisable template with which to explore individual cities and in turn interrogate the usefulness of such concepts.

And finally, one further way in which cities have been classified provides an important way in which to contextualise the remainder of this book. Savage and Warde (1993: 39–40) argue that the spatiality of urban life can broadly be categorised in five specific ways: global cities; cities in developing countries; cities in former socialist countries; old former industrial cities; and new industrial districts. While this list is not comprehensive and, again, cities fall into more than one category, this schema provides a useful way of highlighting that the literature, debates and case studies that are found throughout this book are predominantly focused on theoretical and empirical research undertaken in particular contexts. Given the dominance of consumer culture in particular countries and cities on the one hand, and the uneven proliferation in other countries on the other hand, this book focuses predominantly on the former – western, global cities, new industrial spaces and older industrialised cities.

## Consumption: general and specific

What is vital to grasp from theoretical definitions of the increasing importance of the relationship between consumption and urban change is that both are mediated via a complex interaction of political, economic, social, cultural and spatial practices and processes. For example, at a macro level, consumption is affected by governmental policy and law-making by supranational (e.g. European Union (EU), North America, including Canada, the USA and Mexico (NAFTA), South-East Asia (ASEAN) and Arab, African and Latin American countries (MERCOSUR)), national, regional and local authorities. Uneven economic development and the workings of an increasingly global economy have ensured that the spread of both urban development and consumer culture is itself uneven. Moreover, the social construction of class, gender, sexuality, ethnicity and age, for example, impact on the differential and discursive nature of how, when and what we consume (Clarke 2003).

Consumption is clearly a key area of social life in which tensions and structures are manifest on an everyday basis. Miles (1998b) describes this as a 'consuming paradox', arguing that consumption is both enabling and constraining. Stressing the ideological dimension of consumption, Miles (1998b: 70) argues that 'we play out a lot of the frustrations of contemporary social life through consumption, which gives us some sort of control and certainty to our lives, alongside concomitant feeling that through consumption we can be ourselves'. Moreover, while consumption has become the dominant organisational focus for our social lives, it also ensures that consumers are never entirely satisfied, always striving to consume more (Campbell 1995). Nevertheless, while this indeed appears to be a pervasive ideology, we must not fall into the trap of thinking it is a one-way process. Consumers are not just passive recipients of goods and services; neither can they buy images and identities off the shelf and adorn them at will. There are power dimensions that underlie what, where and how we consume, and people are at least partially aware of such discourses of expression and restriction.

Thus, while consumers are well adept at endowing things with their meanings, making consumption a personal signature and actively using consumption as a resource through which our identities and lifestyles are played out, there are pervasive limits. This is the 'consuming paradox' – we personalise the impersonal. However, while consumption seems to offer everything, there are powerful economic, social, cultural and even spatial restrictions. We can only consume what is on offer, what we can afford (or what our credit limit can stretch to), and moreover we are constrained by legal, social and cultural conventions. As such:

> Consumer capitalism actively wants consumers to experience what might be described as a 'pseudo-sovereignty'. The individual's experience of consumerism is thus a balancing act between structure and agency. . . . The consumer is offered a veneer of sovereignty and maximises his or her personal

freedom within the veneer provided, despite a tacit acceptance that consumerism is a more powerful beast than any one individual at any one time.

(Miles 1998b: 156)

With such commentary in mind, it is clear that consumption expresses generalisable yet unique aspects of urban life. This is precisely why studies of the
relationship between consumption and urban spaces/places and the identities,
lifestyles and forms of sociability that constitute them have made such an
important contribution to making sense of the city.

However, it is clear that Miles's (1998a) contention that, to the individual,
consumption is both enabling (in terms of personal fulfilment and construction
of sovereignty) and constraining (as it plays an ideological role in constructing
lives) can equally be applicable to cities. This 'consuming paradox' suggests that
cities are not helpless pawns reacting to global processes. However, their reactions
are both related to sedimented local and regional political, economic, physical
and social opportunities and conditions, and highly relational in terms of the
activities of other cities.

For example, increasing global interdependence has ensured that the most
economically successful cities *are* those being reconstructed and sold as centres
not of production but of consumption. Important to this are new patterns
of regulating space, which reflect the socially selective location of highly valued
economic activities (Zukin 1982). As such, economic growth is dependent on
cities' abilities to innovate and to produce qualitative and quantitative consumption opportunities. Consumption is thus intrinsically linked to quality of life,
and consumption is the key symbolic content (Lash and Urry 1994). Hence, as
cities strive (with differing degrees of success) to be consumption centres, the
quantitative and qualitative differences between urban landscapes come to the
fore.

In order to successfully address these differences there is a need to focus on urban
life as a complex mediation, the interpretation of which demands analysis of
multiple and interrelated foci – from the nature of the political economy, the social
relations which enable the production of consumption, and the specificities
of consumption cultures and activities (Crewe and Lowe 1995). In sum, this
importantly demands a mesh of both production-oriented study of urban change
*and* culturally derived understandings of consumption and identity. To date,
however, there has overwhelmingly been a focus on highlighting generalisable
phenomena with which to describe the (post-industrial/postFordist/postmodern
nature of cities and regions. In studying factors such as the social relations of
production and the discursive construction of consumption practices, research has
often failed to adequately consider how space and place make a difference to local
practices and processes (Crewe and Lowe 1995).

In terms of urban studies, this has been manifest in a (slow) movement from descriptions of 'ideal' or 'archetypal' social groups, towards a focus on particular modalities of consumption; and the proliferation of literature which seeks to ground consumption practices, lifestyles and identities in specific settings and spatial environments. Such approaches seek to identify historical shifts as well as continuities. This importantly includes an acknowledgement that space and place are not passive backdrops to human relations, but are active in political and economic formations, social relations, subjectivity and social selfhood (Zukin 1991, 1998a; Mort 1996, 1998; Wynne and O'Connor 1998). The relationships that exist between people and places, and how the political, economic, social and cultural practices and processes that mediate the nature of specific cities and the lives of their inhabitants through time and space, need to be considered.

These advances have been important in two ways. First, there has been an increasing recognition of the need to move away from the 'tyranny' of studying single sites such as malls, department stores and urban villages (Jackson and Thrift 1995). This was born from an understanding that if studies of the geographies of consumption were more fully to theorise urban change then the complex 'tissue of sites' which constitute and connect localities, cities, regions, nations, as well as supranational and global spatiality, must be scrutinised (Crewe and Lowe 1995). Secondly, and intimately related to this agenda, was an understanding that political, economic and cultural events are 'embedded' in particular localised social systems (Jackson and Thrift 1995). As such, lifestyles, identities and social relations must be understood as being differentially and actively constructed and negotiated in specific places and spaces, which give them their own 'unique' characteristics. The kind of understanding that such perspectives promote is suggested here to be an important advance. While gaining from the findings of structurally biased research, such an approach presents real potential to enable interpretation of specifically located social and cultural practices and processes and provide insight into the nature of uneven urban development. There is a need to look at how the consumption cultures of specific local contexts can inform our understanding of their broader urban agendas, through the way they either mirror or differ from such practices and processes.

## Consumption, production, representation, identity and regulation

What should have become clear to readers over the course of this introduction is that consumption is not a topic that can be studied in isolation. One way in which its connections have been conceptualised is in terms of the 'circuit of culture' – a model that arranged production, consumption, identity, representation and regulation in a loop to describe an endless circuit (Du Gay 1997). This is important, first, as it links the public and private spheres of consumption, as well as ideological structures and personal agency. Secondly, it identifies that objects of consumption are 'texts' which can be 'read', interpreted, appropriated and re-appropriated, thus stressing the interpretive and productive nature of consumption. This interpretive action is then fed back into lived culture and thus informs the ways that everyday life is played out, an articulation that then provides information upon which production acts. Thus, thirdly, consumption is seen as both articulating and enacting social relations and human action.

It is clear from Figure 1.2 that this model is useful and can be used to discuss consumption in a variety of ways (including, for example, the interaction between objects, images and spaces/places). The ways in which consumption relates to production (including component elements such as social relations) and interpenetrates with regulation (including government legislation), representation (including, for example, media images, place promotion, or popular cultural forms) and identity (individual, social groups and spaces/places) will be utilised to answer questions addressed in this book – how are cities moulded by consumption, and how is consumption moulded by cities?

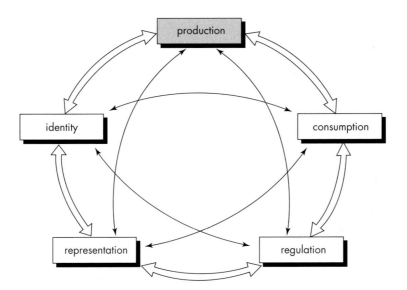

**Figure 1.2** *The circuit of culture. (Courtesy of Sage Publications Ltd)*
*Source: Du Gay (1997)*

## Consuming this book

The next two chapters of the book will identify the relationship between consumption and the development of the modern and postmodern city. The chapters will explain these perspectives by engaging in more depth with the genealogy of the literature outlined in this introduction. For example, Chapter 2 will review the work of early writers, including Veblen and Simmel, and the ways they depicted consumption as a marker of social prestige. The Frankfurt School thesis will be further discussed, as will Marxist depictions of the role of *collective consumption* in terms of agendas of equality and social justice. It will be shown how consumption cultures were responsible for stimulating the development of spectacular spaces/places within the modern city. The main aim is to critically trace how the development of the modern city can be characterised by the relationship between consumption and archetypal identities, lifestyles, forms of sociability and urban spaces and places.

Chapter 3 will show how consumption cultures were responsible for stimulating the development of spectacular spaces/places within the postmodern city. As with Chapter 2, there will be a focus on the relationship between consumption and archetypal identities, lifestyles, forms of sociability and urban spaces and places. This will be contextualised through literature that has presented a new stage of consumerism, based on *individualised consumption*, and its symbolic and aesthetic relation to the emergence of a new middle class will be discussed. Chapters 2 and 3 will be framed though an unpacking of the separate, yet overlapping conceptualisation of post-industrial, postFordist and postmodern urbanity.

Chapter 4 addresses the relationship between consumption and everyday life. This chapter acts as an important antidote to the spectacular urbanism outlined in the previous chapters by focusing on 'ordinary' and 'mundane' consumption, and 'everyday' spaces, activities and social relations. Studies of *inconspicuous* consumption spaces such as car boot sales, charity shops and retro/second-hand clothes shops, markets and consumption within the home will be combined with alternative spaces of consumption in our cities. As such, mundane and everyday worlds, often knowing and transgressive, are shown to have hidden codes and languages every bit as exotic as spectacular spaces, places and formal retailing (Crewe 2000). Moreover, the ways in which people engage with the spectacular urban landscapes in ordinary and mundane ways will be highlighted. This chapter will show the ways in which the individual agency relates to consumption practices, highlighting the weakness in more structurally biased urban studies of archetypal urban spaces/places and identities, lifestyles and forms of sociability.

Chapter 5 looks in more detail at the relationship between cities, consumption and identity. This chapter addresses the ways in which class, gender, ethnicity and sexuality have been central to the development of the city. Most work on consumption and the city has focused on middle-class consumption cultures and failed to discuss consumption by the working-class and urban poor (Bauman 1998). This chapter discusses middle-class consumption cultures, being poor in a consumer society, and how extremes of 'haves' and 'have-nots' are written on to the urban landscape. The role of gendered consumption, ethnicity, sexuality and subcultures will also be discussed.

Chapter 6 will show that cities, and spaces and places within them, not only are sites of consumption but are also themselves consumed. This chapter discusses representations of the city in the 'virtual world' of consumer society – in films, books, magazines, advertising, fashion and songs. Moreover, the ways in which the city is also increasingly represented in place promotion, planning and other official discourses will be discussed. The work of Urry (1995) will also be used to show how we visually consume spaces and places, as well as through other senses such as smell, sound and touch – all of which have an effect on how we interpret and experience the city (Fyfe 1998). Finally, the work of Lefebvre (1991) and Soja (1996) will be discussed in order to investigate the relationship between the 'imagined' and the 'real' city. This will highlight that people consume urban spaces and places in relation to a complex matrix of identity positions but, moreover, that consumption of the 'imagined' city informs the material development of our cities.

Chapter 7 will discuss consumption and urban regeneration, and argue that in a global urban hierarchy characterised by intense competition, cities are promoted and sold not simply as centres of economic growth but as culturally rich places in which to live and work, where the quality and quantity of consumption opportunities are crucial elements in generating place myths. The chapter will identify that central to cities' attempts to move away from a dominance of industrial production to a post-industrial 'knowledge economy' has been the development of a political economy which supports the marginalisation of the industrial past, as well as a cultural and creative economy which enhances a city's liveability. This chapter will critique the seemingly ubiquitous presence of consumption-led urban regeneration initiatives and the implications for cities throughout the urban hierarchy.

Chapter 8 concludes the book by briefly outlining and summarising the key themes covered in the book, revisiting important debates and case studies drawing links across the chapters. It also highlights a future research agenda for studies of cities and consumption.

> **Learning outcomes**
>
> ● To be critically aware of different theories of consumption
> ● To have an appreciation of the relationship between consumption, production, regulation, identity and representation
> ● To be aware of different definitions of the city
> ● To understand that consumption cultures are historically and spatially diverse

## Further reading

Robert Bocock (1993) *Consumption*, London: Routledge. An accessible introduction to the ways in which consumption has been considered by different academic disciplines, which provides rich detail and useful examples to illustrate complex arguments.

Pierre Bourdieu (1984) *Distinction: A Social Critique of the Judgement of Taste*, London: Routledge and Kegan Paul. A seminal book that has been centre stage in debate concerning the proliferation of consumer culture. Blending theoretical debate and empirical evidence to great effect, this book is a vital contribution to students of consumption.

Paul Du Gay (ed.) (1997) *Production of Culture/Cultures of Production*, London: Sage. An important collection of essays by leading scholars who draw connections between the production, consumption, representation and regulation of cultural forms and practices.

Steven Miles (1998) *Consumerism as a Way of Life*, London: Sage. A clear, concise and student-friendly overview of studies of consumption. Explaining complex theoretical arguments in an accessible manner, this book is a useful primer for those interested in the broad and varied topic of consumption.

# 2 Consumption and the modern city

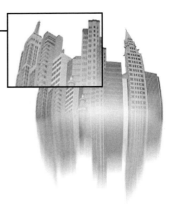

**Learning objectives**

- To look at the relationship between consumption and the development of the modern city
- To describe the new archetypal urban spaces and places associated with consumption cultures
- To think about the relationship between production and consumption and the social control of the city
- To describe the relationship between new archetypal identities, lifestyles and forms of sociability, consumption and the modern city

This chapter will show how consumption was inextricably bound up with the development of modernity and hence its most potent symbol of progress and achievement – the modern city. It is in the modern city that consumer culture and the core political, economic, social and cultural institutions, organisations, and the infrastructure of modern life are brought together to a degree and extent never seen before. For example, the strengthening and formalisation of local municipal government power, planning regimes and service provision emerged from the political and economic dominance of industrialists over the development of the modern city. Similarly, the expansion and increasing internationalisation of stock, goods exchanges and markets ensured that cities were increasingly a nexus point for flows of money, goods and people from all over the world. This economic, social and cultural milieu was aligned with, and was integral to, the rapid proliferation of shops, banks, sports clubs, cinemas, theatres, pubs, cafés, restaurants and a mix of urban dwellers from an increasingly diverse range of social and ethnic groups. It is this complementary strengthening of political institutions and organisations and economic growth (and a parallel development of cultural and consumption-oriented activities) that dramatically impacted on the development of the modern cityscape.

The relationship between consumption, modernity and the development of the modern city is founded on rational organisation, planning, science and technological advances. Equally important is the understanding that it is in the modern city that consumption became bound up with the very idea of what the modern world was about, in particular how individuals experience and view selfhood and identity. For example, Slater (1997) argues that it is through the development of the modern city that individuals were no longer governed by tradition but by constant change and flux, and that it was the growth of consumer culture that was integral to making such conditions in the modern world. Consumer culture is therefore not simply a product that materialised after the industrial and intellectual success of modern thought were accomplished – but rather 'the consumer' and the experience of consumerism were integral to making the modern city.

## Archetypal structural and socio-spatial transformations of the modern city

Given this close relationship, it is not surprising that analyses of the growth of consumer culture and the physical, political, economic and social transformations of the modern city have been prominent in urban studies since the beginning of the nineteenth century. From that time onwards, theorists have provided accounts of how social groups are active in effecting urban change. Their work has often generated 'archetypal' descriptions showing how certain groups gained dominance over particular urban spaces and places. Such urban theory has played an important role in linking consumption to structural and institutional change, as well as to sociability and changing social relations. It is through key topographical landmarks of the modern city that analysis has charted this relationship, through studies of arcades and department stores, the spectacular settings of large-scale redevelopments of boulevards, squares and public spaces, and the archetypal social groups associated with their development. What follows in the rest of this chapter is an introduction to the literature, places, people and sights that constitute a genealogy of the development of the modern city. This review allows us to unpack the relationship between the birth of consumer culture and the development of the modern city.

## The birth of the modern city (1880–1930)

McKendrick *et al.* (1982) suggest that antecedents of the consumer revolution emerge for the first time in the UK during the eighteenth century. They argue that for the first time a society developed within which material possessions were prized less for their durability and more and more for their fashionability. It must be noted, however, that this early consumer economy emerged from a longer historical tradition. For example, during the seventeenth century, markets, fairs and carnivals became focal points for consumption and leisure throughout Europe, and later, in the eighteenth century, small-scale producers emerged and profited from demand for consumer goods such as pottery, jewellery, clothing, buttons and pins. However, McKendrick *et al.* argue that such early production and consumption economies rapidly grew in importance during the eighteenth century. Moreover, such rapid development occurred predominantly in urban areas, which at that time began to be increasingly influenced (via the expanding networks of canals and roads) by the economic and cultural activity of strong European nations and the emergence and growth of capital cities.

For example, Peter Corrigan (1997) notes that conspicuous and flamboyant consumption cultures can historically be associated with the fashion-conscious European aristocracy. Corrigan highlights that during the Elizabethan period 'fashion' became an important concept to the aristocracy, and in particular London became a centre of elite consumption. It was thus in larger cities that commerce, traders and larger populations began to generate not only demand and supply for consumer goods but also a proliferation of shops of many kinds, advertising, window displays and urban leisure consumption in music halls and pubs. As such, the development of urban life has always been the product of, and the catalyst for, the consumer ethic and the birth of consumerism.

However, it was through rapid industrialisation in the eighteenth century, and an associated rise of a mass production and mass participation society, that the seeds of consumer society sown over earlier periods of history fully blossomed. It is at this time that changes in production techniques took place hand in hand with changes in people's tastes, preferences and desires. For example, in 1899, Thorstein Veblen, the first sociologist to describe the importance of consumption in the modern life, depicted the increasing importance of consumer goods as markers of social prestige and status. Veblen identified the rise of the *nouveaux riches* of late nineteenth-century America as an important social group who drew on their ability to buy goods and services in order to show their wealth and success. Veblen (1899) shows that this had much to do with a mimicking by this 'new leisure class' (who had grown wealthy from the economic industrial revolution through coal, steel, textiles and other industrial production) of the consumption practices of the upper classes of Europe.

However, in the light of the flamboyant consumption activities of the *nouveaux riches* (and the parading of the new-found wealth based on the successes of the industrial revolution and the resultant explosion in finance and banking), higher social groups responded by continually updating their consumption practices in order to stay ahead of this new social stratum. As such, consumption became increasingly significant in its role, as it conferred status and thus increasingly played an instrumental part in underpinning an emerging new social hierarchy. With the social structure increasingly being based on the high-profile consumption practices, Veblen identified an elaborate system of rank and grade which marked through consumption a person's place in the social world and class hierarchies. This is a process called 'pecuniary emulation', and Veblen was correct to identify the extent to which consumption was having a growing significance in the construction of everyday life.

It was not until twenty years later that Georg Simmel (1957) took up the lead given by Veblen and further investigated the increasing significance of consumption to the construction of modern social life at the beginning of the twentieth century. In *Metropolis and Mental Life*, Simmel described the importance of money and specifically exchange as a central influence on modernity. He argued that the growth of political and economic organisations and institutions in the modern metropolis meant that it was the agglomeration of economic, social and cultural practices and processes in the city that drove the development of a mature money economy.

Importantly, Simmel argued that underpinning the growth of the urban economy was a proliferation of social relations that were increasingly anonymous – less based on personal relationships but rather on market-based relationships. The growth of the city had affected conduct, social life and social relations – the way people engaged with and perceived their fellow city dwellers. Simmel showed that it was through consumption that people learned to negotiate and cope with these changes. He highlighted that as the city grew, consumption began to satisfy the social and psychological needs of its inhabitants and that fashion as a form of social demarcation was becoming increasingly important.

It was through fashion that individuals could allay feelings of insecurity and compensate for the awe-inspiring and overwhelming effect that the sheer size of the city and the numbers of people – all of whom were strangers – were having on urban dwellers. By both asserting and communicating individuality and inspiring feelings of security and commonality (by associating with the fashion choices of others), urban dwellers – but in particular the *nouveaux riches* – could attempt to identify and mimic their social betters, at the same time as distinguishing themselves from the working classes. This represents a contradiction between 'belonging to' and 'separation from' a social group, and also represents a need for individuation and distinction from the members of that group. Thus, in the increasingly commercialised spaces and places of the city – and as the fast pace of urban life became more and more intense – it is fashion, Simmel argues, that provides a means of grounding oneself, of stabilising the assault on the senses that characterises the flux and ever-changing 'buzz' of the city. It is clear, then, that in the writings of Veblen and Simmel consumption is acknowledged as having an important role in the social and cultural everyday life of the city, acting as a buffer between individuals and the everyday tensions which characterised the early modern city for its inhabitants.

It is important to note that the development of the modern city was not only being organised around shopping spaces and places. The development of the early modern city also engendered a variety of public policies, institutions and governance mechanisms intended to mitigate the failures of the market, and to reform modern industrial arrangements and practices. It was left for local authorities to provide the infrastructure and service provision that the free market at this time did not deem profit-worthy – in sum, providing the support for urban growth, workers and factories such as utilities, sewerage, poor relief, refuse collection, education and transport.

In order to unpack how the relationship between collective consumption and consumer culture was theorised it is important to begin with the work of Karl Marx. While Marx was very much involved in theorising the role of production in capitalist societies, the concept of the commodity is a springboard for many theories of consumption – in particular, what constitutes human needs and how those needs are met through the production process. Marx sees the object of labour (that is, the material artefact or product) as having a crucial role in the construction of people's lives and, in turn, in their sense of well-being. In these terms, Marx understood that the commodity is contextualised purely by the production process. It was production rather than consumption on which he focused his argument. Marx describes a commodity as a product that has been produced not for direct personal consumption on the part of the consumer, but rather with the intention of selling it in the marketplace. As we saw in Chapter 1, it is in this sense that the commodity becomes significant for its exchange value rather than simply its use value. The key point is that the value of a commodity appears to be natural or objective, when in fact the actual value of the goods bears no relation to their use value. Rather, this value is dependent upon the concrete social relations of capitalist production, which actively camouflage the exploitative social and labour relations that underlie the process as a whole.

What is useful about Marx's ideas is that he highlights the formative role of the commodity in everyday life. He shows that the workers' relationship to the means of production and thus to the commodity is a key influence on individual experience. Such 'commodification', he argued, meant that all aspects of social life become subject to the laws of the marketplace. As well as goods and services, people and relationships are reduced to a monetary value in an increasingly alienated capitalist world. The legacy of this commodification was the alienation of the worker through the extraction of surplus value. This represents the corner-stone of the argument that workers are 'forced' to become consumers by buying the products they or their fellow workers have made. By earning wages or salaries, and being obliged to become embroiled in consumer culture, the wheels of capitalism are oiled by the workers themselves.

Marx did, however, argue that commodities symbolised the power of capitalism, because the power structures that lay behind the commodity underpinned capitalism. The commodity plays an important role in relating the individual to the capitalist system. However, Marx argued that this created a false consciousness that hides the realities of what he describes as alienation. Commodification, the process by which everything is valued according to its value in the system of exchange, creates a world in which a person's priorities become subject to the requirements of the market. The worker no longer has ownership of the commodity they produce, thus alienating them as a bit-part player. Marx describes this as a fetishism of the commodity, giving it mystical qualities, with significance beyond its use value – to be treated with awe and reverence. Marx considered that it is production and people's relationship with the means of production that is a key determinant of social structures and relationships. Marx viewed the commodity as something to be sold and exchanged and therefore as having a role in determining your social position, but considered that that position is not determined by how you actively engaged with consumer goods.

It was the Frankfurt School for Social Research, set up in 1922 and founded mainly by left-wing middle-class German Jewish intellectuals, that initially took up Marx's intellectual legacy. The Frankfurt School moved to America following the rise of the Nazi party in Germany, moved back to Frankfurt in 1949 and was disbanded in 1969. Members of the School used Marx's theory of commodity fetishism to show how cultural forms function to secure continuing economic, political and ideological dimensions of capital.

Writers such as Theodor Adorno, Max Horkheimer and Herbert Marcuse (the work of Walter Benjamin will be discussed later in this chapter) worked on Marx's concept of alienation and the way this emphasised how the mass production of cultural artefacts such as music and literature produced a 'culture industry' that integrated workers into the capitalist society. According to mass-culture theory, the 'genuine' folk culture of the people was destroyed by a highly commercialised and standardised mass culture, which also undermined the standards of high culture. Adorno and Horkheimer's term 'culture industry' describes the transformation of culture. Culture had become something that was made and sold, just like any other industrial product, in order to make a profit. It was, according to this view, imposed on the masses by the culture industry and turned people into passive consumers of material that did not meet their real needs. Mass culture was crucial to maintaining a capitalist society. Workers were willing to accept boredom and exploitation at work because they could escape during their leisure hours into the pleasure of popular culture, shopping, watching films and listening to jazz.

Following Simmel, Veblen and Marx, a very influential thinker who has illuminated our understanding of the impact of consumption was Pierre Bourdieu. Although writing much later than his predecessors, in the 1960s and 1970s, Bourdieu advanced Simmel's and Veblen's work and utilised empirical research that looked at patterns of consumption in France. Bourdieu showed how the social significance of consumption is not simply in expressing variations between different social groups but in having a formative role in establishing those differences. Consumption is thus a cultural resource that underpins our everyday lives.

In *Distinction*, Bourdieu (1984) argued that we are motivated by a need to establish, protect and reproduce a collective pattern of preferences based on class difference, and that different classes are differentially 'educated' to take advantage of their symbolic reading of the world. It is through possession of 'cultural capital' that dominant classes demonstrate their superiority via access to high culture and high consumption. Of course, access to economic resources is central to acquiring cultural resources, but Bourdieu argued that it is the social and cultural *norm* for the individual to aspire to accruing cultural capital. It is through consumption that social groups can mark themselves from one another – and the 'cultural capital' accumulated through consumption is formulated by signs, symbols, ideas and values.

A central element to Bourdieu's thinking is that consumption habits are not simply the product of social structures, but rather are an interaction between individuals and society. In order to explain this he developed the concept of 'habitus' – the everyday know-how we gain from routine experience and how we learn 'appropriate' behaviour. Bourdieu argued that habitus underpins power relations based on a system of perceived difference between social groups. This difference is constructed through, and can be read by, things like our identities, lifestyles and fashions, which delineate what is distinguished as vulgar, right or wrong, cool or uncool, to different groups.

Habitus thus allows group distinction, providing a framework for social recognition, understanding and interpretation that is reproduced between generations and also generates the schemes by which cultural objects are classified and differentiated. As such, class differences are inscribed on individuals as differences in taste. In sum, our social experience is structured by what we see as being legitimate ways to do things according to classifications of taste. In these terms, Bourdieu thus provides a significant insight into the construction of class difference and taste that has been central to the development of the modern city.

Integral to the growth of this social and cultural role was the way in which, in time, the burgeoning consumer culture was written on to the physical landscape of the modern city. For example, as the importance of fashion grew, so did the spaces and places of consumer culture where fashion could be viewed and purchased, and it was during this period that there was a proliferation of shopping arcades, department stores, cafés, theatres, restaurants and other new forms of entertainment such as cinema and professional sport. However, it is important to note not only that consumption was important to the progress of cities, but that cities themselves were central to showing progress of modern nationhood. Theatres of modernity, such as museums, theatres, bridges, roads, department stores and especially international exhibitions, were powerful symbols of scientific progress, civil success and national greatness. The growth of the modern city went hand in hand with colonialism, the firming up of national identity and centralised political power in western industrialised countries. This national pride was celebrated at events such as the Great Exhibition in London in 1851 and the Paris Exhibition of 1889 where all that was great about technological, scientific, economic and cultural progress was put on show.

The world of consumable experience and goods was delivered by modern progress into a modern carnival present in the sights, sounds, and space and places of the city, and the consumer became the fee-paying audience of the experience and spectacle of modernity. It was at this time that consumer culture was underpinned by continuous self-creation and accessibility to things that were presented as new and fashionable, always improved and improving. It was in cities that this speeding up of consumer culture was most explicit. Before looking in more detail at consumption spaces and places, the following section will unpack how consumption and production relate to social control and the physical development of the city.

## Social control and the city: the growth of consumer capitalism

Public cultures of consumerism were being made fashionable and respectable during the growth of the early modern city through their connection to the *nouveaux riches* and their public purchase and display of fashion and construction of private bourgeois domesticity. For example, Siegfried Kracauer (1926) describes this as a process of 'mass ornamentation', where symmetry, rationalisation and decoration of public spaces proliferated into the domestic sphere. Kracauer pointed to the influence of bastions of modernity such as theatres, hotels, restaurants and government buildings which mimicked the elegance of stately homes, and in turn influenced the way in which the new middle classes decorated their homes. However, while consumption had been turned into a 'respectable' culture by wresting it from the worst excesses of the aristocracy (where it signified luxury, decadence and superficiality), the consumption cultures of the *nouveaux riches* were also important in terms of distinction from, and imposing control over, the working classes (amongst whom consumption cultures were based around public riotousness and the excesses of drinking and blood sports). During this period much of the public debate on consumer culture, in particular in newspapers, public meetings, parliament and local government, was carried out not only in terms of consumption of goods but also in terms of their relationship to time and activity. This debate centred on the leisure pursuits of the working class – and the big question for the bourgeoisie was how to keep public order outside work hours and indeed how to improve the productivity of the working classes during work hours (Holliday and Jayne 2000).

How, for example, could working-class consumption patterns of excessive drinking, blood sports and prostitution be eradicated from public places – and even if they could, there was concern about what the working classes got up to in private. Such considerations were characteristic of middle-class fears for health and morals, and the subversion and irreligion of the working class became central to political debate during this period. The rise of Victorian philanthropy and political reform was very much underpinned by the wish to impose a particular set of consumption activities – of healthy, domestic, private consumption in the unit of the family – and to produce pious, productive and sober workers. This was focused on a whole range of initiatives – from the provision of libraries and museums, and the building of public bath houses, municipal sewerage systems and electricity and gas supplies, to the promotion of 'rational recreation' for working-class leisure time such as sports and woodwork. This raft of urban reform and infrastructure was pursued by the industrialists who dominated urban political power and sought to exert control over their workers in order to improve their profitability. In sum, consumer culture in the mid-nineteenth century appears to emerge from a series of struggles to tame, yet at the same time exploit commercially, the social spaces and times in which modernity is acted out (Slater 1997).

These concerns and conflicts were themselves manifest in the physical development of the modern city. For example, Malcolm Miles (2004) describes the archetypal process of redevelopment that took place in Paris from 1853 to 1870, which was repeated in cities throughout Europe. Baron Haussmann, who was the Prefect of the Seine, working to Napoleon III's sketch of how the King thought his great city should appear, was responsible for the redevelopment of Paris. With little concern for the concept of social justice, Haussmann used tall wooden towers to survey the existing urban fabric before measuring out wide streets through the old working-class quarters. The aim was to divide them, while to a large extent displacing the poor out of the inner city to the periphery.

Paris was transformed, as were many other nineteenth-century cities, from a city that was chaotically structured, with narrow medieval streets, into a city with wide boulevards and great avenues. Baron Haussmann redesigned the city not only in terms of aesthetic sensibilities but in order to better control the unruly working classes. The new wide streets allowed troops to quell popular uprisings more easily and were wide enough for two army wagons to roll abreast (Miles 2004). They linked the inner city to the army barracks and hence allowed city authorities to secure the city against popular uprising. This physical redevelopment took place at a time when municipal authorities were also strengthening control of working-class areas through the inception of police forces.

Importantly, this redevelopment of the city led to new ways of consuming the city. There were new urban spaces that created sights and views that could not be seen in the old Paris. The great buildings and boulevards became places where large numbers of people circulated, and people became part of the sight of the city. Hausmann's plan included the building of markets, bridges, parks, the opera and other cultural palaces, with many located at the end of boulevards. For the first time in a major city, people could see well into the distance and indeed see where they were going and where they had come from. Great sweeping vistas were designed so that each walk led to a dramatic climax (Urry 1990). The boulevards made walking in the streets more visible, offering opportunities to be seen while affirming an increasingly restrictive climate for women's use of streets.

For the poor, too, Paris became a city of exclusion as areas of informal housing were cleared. This led to a new problem of vagrancy to which Haussmann, largely responsible for, reacted with vitriol. The redevelopment was a brutal redefining of the city and bourgeois ideals of urbanity, an attempt to expel the working class from central areas of the city. This included an inherent reformist zeal to improve the lives of the working class to make them more pious but, importantly, in terms of the development of capitalist accumulation, to improve productivity.

For instance, one further vital function of the redevelopment of medieval cities was the separating out of industry from home, resulting in zoning in which industry and recreation were separated from domestic life and shopping, and neighbourhoods differentiated by class (Pile *et al.* 1999). This was as much to do with controlling and policing production and consumption cultures in specific locations, and was supported by the burgeoning public transport system to take people from work to home and to central shopping areas. This allowed space for factories to develop and central areas to be free to expand without the constraints of proximity to working-class slums. In pre-capitalist society the social relations between work and home, work and season, town and country, dictated a particular regime of accumulation. This accumulation regime changed with the birth of the modern industrial city and was a transformation not merely of the labour process. For example, one important element was that the estrangement and objectivity of labour were intensified by the systems of rational planning and administration which gathered pace under advanced capitalism. The redevelopment of Paris provided a blueprint for the modern city – one that was implemented in 'emerging and expanding cities in every corner of the world, from Santiago to Saigon' (Berman 1992: 152).

## The city as machine

Integral to the global rise of modern urbanity was the emergence out of the industrial revolution of Fordist production techniques. Fordism refers to the system of mass production and consumption characteristic of developed economies during the 1940s–1960s. Under Fordism, mass consumption combined with mass production to produce sustained economic growth and widespread material advancement. The rise of Henry Ford and the Ford Motor Company was a symbol of the complete transformation from an agricultural to an industrial, mass-production, mass-consumption economy.

Central to this transformation was the combination of new technology, techniques and products. Together these fuelled the industrial revolution and, once rationalised through Fordist production, the movement from craft-based production to industrial capitalism and mass production was completed. It was at this time that markets became increasingly internationalised, based on economies of scale which gave rise to giant organisations, and built upon functional specialisation and minute divisions of labour. It also engendered a variety of public policies, institutions, and governance mechanisms intended to mitigate the failures of the market, and to reform modern industrial arrangements and practices (see Case study 2.1).

Ford's main contributions to mass production/consumption were in the realm of process engineering. The hallmark of his system was standardisation – standardised components, standardised manufacturing processes, and a simple, easy-to-manufacture (and repair) standard product. Standardisation required nearly perfect interchangeability of parts. To achieve interchangeability, Ford exploited advances in machine tools and gauging systems. These innovations made possible the moving (or continuous) assembly line, in which each assembler performed a single, repetitive task. Ford was also one of the first to realise the potential of the electric motor to reconfigure workflow. Machines that had previously been arrayed about a central power source could now be placed on the assembly line, thereby dramatically increasing throughput. The moving assembly line was first implemented at Ford's Model T Plant at Highland Park, Michigan, in 1914, increasing labour productivity tenfold and permitting stunning price cuts. Hence, Fordism was about the standardisation of a product and manufacturing it by mass means at a price low enough for the 'common person' to afford. Ford made everything he needed to produce and sell his cars, from the production of raw materials and components to advertising and marketing. Of course, total integration of the production system required the organisation of huge numbers of activities and employees. The production-line technique required employees completing different jobs with varying degrees of prestige and pay, from shop-floor workers to staff specialists, and, importantly, a whole new tier of middle managers.

One further contribution to the success of consumer capitalism at this time was the way in which every movement of the worker in carrying out a task is examined and analysed to eliminate wasted effort and time, and the formalisation of the resulting analysis as a productivity norm that each worker must reach. Frederick Taylor wrote *The Principles of Scientific Management* in 1911, and these principles became known as Taylorism. The principal object of management should be to secure maximum prosperity for the employer, coupled with maximum prosperity for the employee. The corporate order, with its assembly-line techniques, job differentiation and increased organisational size, demanded a different type of factory and office space and a more regulated and regimented flow of time and work.

## Case study 2.1 **Collective consumption**

Theorists have drawn on the intellectual genealogy from Marx and the Frankfurt School in order link individual agency, lifestyle and forms of sociability to structural and urban change. The conceptualisation of *collective consumption* was described by Castells (1977) and Saunders (1981). This conception, embedded in a neo-Marxist approach, sought to describe processes of collective consumption as a definite characteristic of the city under advanced capitalism. At this time they considered that the city achieves its distinctiveness as a result of the way in which the reproduction of labour power operates through, and simultaneously structures, space in a manner directly related to the means of production. In essence, collective consumption became a functional necessity for the reproduction of capital.

For the city to become economically successful, and function so that workers could enable capitalists to maximise profits, housing, gas, electricity, water, sewerage, education and transport had to be organised around both the needs of industrial production and the domestic needs of workers' and owners' homes. As such, local authorities, heavily influenced by industrialists, planned and developed cities around the public provision (through levying taxation) of and collective access to, and hence consumption of, services and utilities that were vital to ensuring the smooth running of industrial capitalism. This was also structured in social space through segregated production and consumption practices – relating to the workplace and other institutions such as family, home, church, schools, and consumption places and spaces – and was underpinned by fixed-identity categories of class, gender and ethnicity.

This emphasis on consumption studies initially gained momentum in the early work of Manuel Castells (1977, 1978), David Harvey (1973) and later the ideas of Peter Saunders (1981). These theorists sought to depict the role of *collective consumption* (such as housing and other service provision) in terms of agendas of social justice. For example, writing during the 1970s, when a series of crises affected capitalism as a whole, Castells (as noted in Chapter 1) described a situation where the decline of traditional heavy industry and manufacturing, and the growth of new high-tech and service-oriented employment, was leading to massive loss of jobs. Castells argued that the effect of such de-industrialisation and loss of wealth and revenue was leading to a reduction in city authorities' ability to pay for welfare and services via taxation. As such, spending on infrastructure and services would be reduced as local government sought to balance budgets.

Castells realised that it was collective consumption that would be first hit by such fiscal crises and that a responding popular labour movement's understanding of workers' subordinate position in relation to the means of production would lead

continued

to revolutionary reconstruction. Such a revolution did not, of course, take place, and the concept of public ownership of utilities and services and their collective consumption for the greater good of the city and its inhabitants was being slowly eroded at the time that Castells was writing, during the 1970s – and was jettisoned altogether during the 1980s. The 1980s were dominated by 'neo-liberal' economic policy, espoused by Conservative political parties and their support for individual rights, individual responsibilities and deregulation that led to an erosion of local authority control and tax-raising powers and the privatisation of utilities and services such as housing and transport. The labour movement that Castells saw as key to revolutionary change was also undermined both through the decline and loss of jobs in manufacturing and heavy industry, and also through strict legislation that limited the power of trade unions. It was at this time that at the level of political support and policy-making, individualised consumption predominantly replaced collective consumption as a key agenda in the redevelopment of cities and the postFordist organisation of the capitalist economy (see Chapter 3).

In sum, the domination of consumption through the logic of Fordist production was established in the modern city through the production of standardised goods. Such goods, with a functional aesthetic amenable to automated mass production, could now be sold to the large numbers of people in close proximity to one another in urban areas. As such, Fordist mass production requires a large scale of social consumption – built around economies of scale, therefore decreasing unit costs – but made sustainable by standardising goods. This provided a high wage for workers and new managers that enabled them to buy products. Demand for the product was ensured by advertising campaigns that spread the illusion that consumers could differentiate themselves from their neighbours by purchasing the same standardised goods.

The restructuring of the modern city from the 1880s to the early 1900s and the related economic, social and cultural change laid the foundation that led to the 1920s appearing as the first consumerist decade. This era saw the full emergence of mass production systems of manufacture increasingly dedicated to producing consumer goods (rather than heavy capital goods such as steel, machinery and chemicals, which previously dominated). Importantly, it is in this period that consumer culture took on its mature form, and a modern norm emerged of how consumer goods were to be produced, sold and assimilated into everyday life. Goods were designed with standardised replaceable components which allowed them to be produced in very large volumes at low unit cost through an intensive, rationally controlled and increasingly automated technical division of labour.

The growth and agglomeration of banking, finance and other professional services allowed goods to be sold across geographically and socially wider markets – regional, national and global. The exploitation of these markets was made possible by the interconnection of local markets through new transportation and communication infrastructures (rail, mail, telegraph, telephone) and by the concentration of finance, political power and markets in larger cities. The growth in finance and banking allowed the development of multi-divisional corporations capable of planning and co-ordination, marketing, branding and packaging, co-ordinating national and international sales forces, and undertaking advertising. This was accompanied by the mass development of retail infrastructure (not just shops but also retail chains, and mail order), where the massive volume of cheap standardised goods could be sold through larger and larger markets to a population now seen as consumers.

Figures 2.1 and 2.2 show how the physical organisation of the city is clearly structured around residential and business sectors. These represented a rational use of space and in particular the establishment of distinct zones of production. They also allowed planned provision of collective consumption in both residential and industrial areas – such as heavy industry and manufacturing zones, central business districts, commercial shopping areas, workers villages, other residential districts, and so on. These zones of the city were rationally spatially organised and serviced by industrial and domestic utility provision, transport links and networks, schools, and the everyday consumption needs of different

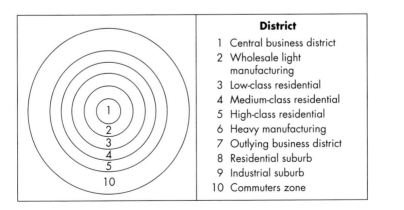

**Figure 2.1** *Burgess's concentric zone model. (Courtesy of the American Academy of Political Science)*

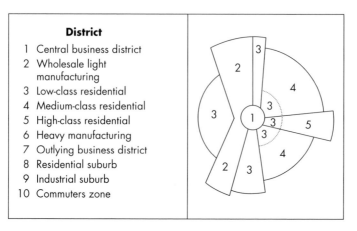

**District**

1 Central business district
2 Wholesale light manufacturing
3 Low-class residential
4 Medium-class residential
5 High-class residential
6 Heavy manufacturing
7 Outlying business district
8 Residential suburb
9 Industrial suburb
10 Commuters zone

**Figure 2.2** *Hoyt's sector model. (Courtesy of the American Academy of Political Science)*

sections of the urban population. While in reality such modelling of the city represented in these two figures failed to account for the more complex and 'messy' organisation of most cities, they do indeed highlight the type of mindset that led to rational master planning of cities around the needs of production and collective consumption.

Integral to the rational planning of the modern city was the emergence of planning and architecture as professional disciplines that were both quickly institutionalised in public-sector urban authorities and utilised by private-sector companies to symbolise economic success. For example, modernism emerged as an identifiable architectural movement and set of design principles during the final decades of the nineteenth century and the first decade of the twentieth. Stevenson (2003) argues that the foundations of the movement lay in Europe, but it was the US mid-western city of Chicago that popularised the new style. Such architecture firmly articulated the physical appearance of the modern city, and impacted on the development of cities around the world.

Modernist architecture put into practice technological and aesthetic innovations that broke sharply with those of the past. Stevenson (2003) argues that it was in Chicago that modernist architectural values were initially and most spectacularly mediated. Chicago built the world's first skyscraper and the city boasted scientifically based architecture and planning regimes. Early modernist architecture was based around the extensive use of glass, iron and reinforced concrete and was focused on a modern concern with rationality, simplicity and efficiency. This resulted in a new architectural aesthetic, with little ornamentation or references to past styles. This articulated faith in technological advance and upheld the modern ideology of progress. Modern buildings were a bold affirmation of scientific knowledge as well as a celebration of new materials and construction techniques. They were a visible and potent symbol of the machine age. Skyscrapers constructed at this time include iconic buildings such as the Empire State Building, the Chrysler Building and the Rockefeller Centre. These buildings, combining height, utility and art deco design, were not just tall and functional – they were beautiful.

By the 1930s, the skyscraper had become a pertinent symbol of modern American architecture, and developed into a truly international architectural style, emerging throughout the world. However, an important movement at the time was the spread of modernist architectural principles from commercial construction into domestic and publicly funded building, and particularly the increasing popularity of concrete. Closely linked to the increasing popularity of modernist principles in municipal construction was the work of Le Corbusier and the Congrès International de l'Architecture Moderne in Paris, which exemplified that modern urban living should be about houses that were 'machines for living'. These machines were based on Fordist models of production that dominated the industrial modern era. From planning to building efficiency, mass production and rationality underpinned Le Corbusier's manifesto – a utopian vision underpinned by the economic, scientific and social optimism of the 1920s. The influence of this philosophy on urban design and architecture throughout the twentieth century was profound, and effected the transformation of cityscapes throughout the globe.

Le Corbusier espoused the principles of linear design, advocated the use of concrete in construction, and developed what he saw as the liberating potential of the 'rational city' through scientific planning and design. The Corbusian city was vertical, and prescribed the number of residents to any one urban zone. Importantly, Le Corbusier considered the skyscraper as the best model for his ideology, allowing land values to be maximised, with free space for parks and gardens. For example, Le Corbusier's 'radial city' was designed to accommodate 3 million people at a density of 1200 inhabitants to the acre. This was a high-density high-rise life that was built according to strict geometric principles, comprising repetitive grid-lines of X-shaped towers. This city was also sharply divided by class, with the workers (who supposedly never went to the city) living in tower blocks located in self-contained 'garden' suburbs on the periphery, while middle-class 'citizens' who worked in the city were housed in high-density accommodation in the centre. These were affluent, cosmopolitan urban dwellers who could afford all that the city had to offer; urbanism for them was about consumption and lifestyle. However, there was no place for street life, as the streets were given over to gas, electricity and water, and cars and other motor

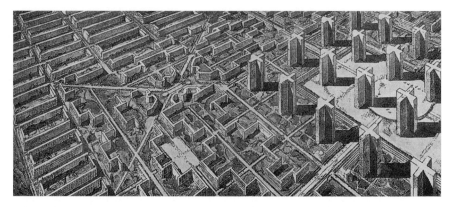

**Figure 2.3** *Le Corbusier's vision of the contemporary city. (Courtesy of FLC/ADAGP, Paris and DACS, London)*

vehicles. Le Corbusier's vision was based on providing collective consumption of housing, utilities, increased social segregation and, for the middle class, the consumption delights of the urban core.

While Le Corbusier and his antecedent, Baron Haussmann, never got to turn their plans fully into reality, their impact on the development of the modern city was profound. What follows in the rest of the chapter is a look at how architecture and modernist urban planning were intertwined with consumption and the people who inhabited the emerging modern city.

## Spaces, places, identities, lifestyles and forms of sociability

The development of the modern city from 1880 to 1930 was dominated by the proliferation of a bourgeois commercial culture that shaped public life and which served the consumption practices of the urban bourgeoisie. This commercialisation of space is seen clearly in the development of arcades and department stores. Other significant urban landmarks included cafés and restaurants, theatres, hotels, public parks, and, particularly in North America, professional sports stadiums and amusement parks. Moreover, from an initial air of exclusivity to upper-class patrons, such places became accessible to lower classes; hence working- and middle-class people mingled openly in the bars and at the racetrack, and in commercial areas (Shields 1991; Zukin 1998b). Similarly, despite attempts to banish the working class from the urban core, cultural districts developed around meeting places and tea rooms, while other nearby districts were more liminal, where illicit and illegal activity often crossed class, racial and sexual lines (Chauncey 1994).

Nevertheless, the most prominent spaces of consumption were the new shopping spaces such as shopping arcades (see Case study 2.2). It was Baudelaire (1955) (followed by Benjamin 1973), who noted the development of these spectacular new shopping spaces and their origins in the arcades of Paris in the early 1800s. Such spaces embodied modern progress – the mass production of consumer goods, technologies of building and display, and in strategies of creating and selling. These spaces were 'bazaars' where previously exotic goods became familiar. Moreover, while such goods had usually been sold only in elegant boutiques, 'the new availability of consumer goods through sight, touch and smell democratised desire' (Zukin 1998a: 827). At this time, then, urban customers could confront these new longings, new freedoms and new incursions into their lives by purchasing 'exotic' things from faraway places, and through occupying the same public space as people who were culturally 'strange' (Leach 1984; Shields 1994; Fritzsche 1996).

## Case study 2.2 **The arcades**

Shopping arcades were high-profile urban spaces that appeared in Paris and other major cities throughout the world in the early nineteenth century. Shopping arcades were architecturally grand spaces designed for new forms of urban consumption. The proliferation of these arcades was underpinned by the growth of metropolitan areas as centres of industry, manufacturing and commerce which enabled an entrepreneurial class to develop and generate a market for luxury goods and sustain a growing middle class that marked itself through conspicuous consumption. Arcades were filled with luxury shops and restaurants that produced a new consumption experience not focused around daily needs, but displaying commodities to be gazed upon, and offering an invitation to locate desire in goods. This 'magical' new world was enabled by architectural design and materials that included large plate glass windows and steel structures, encouraging spectacular window displays. Moreover, the glass roofs of the arcades and the presence of cast-iron structures and gas lamps set aside the arcades from the rest of the city, and created a displacement that signified the arcades as new spaces dedicated to consumption rather than to more diverse practices of urban street life.

The most high-profile writer to consider the economic, social and cultural significance of the arcades was Walter Benjamin, who described arcades as producing 'a world in miniature' – one that engendered 'artificial behaviour'. Describing case studies of the arcades that developed in Paris during the 1820s

continued

and 30s, Benjamin shows that music shops, wine merchants, hosiers, haber-dashers, tailors, boot makers, bookshops and restaurants were usual occupants of the arcades. Benjamin considered the arcades to be dream worlds, a utopian and imagined space, constructed to be better than the real world. Benjamin argued that the arcades thus represented an ideological celebration of the tremendous productive capabilities of capitalism and, while being spectacular, such spaces ultimately made invisible the growing polarisation of wealth in the city. In these terms, the experience of consumption and display of commodities in the setting of urban shopping arcades represented a collective dream under-pinned by promises of a luxurious and relaxed existence.

However, Benjamin argued that central to this utopian vision were 'commodities', and that this relationship between the commodity and the consumer placed the latter in a disempowered position where happiness and well-being were depen-dent on consumption. As such, despite the glimpses of utopian life that were on offer in the arcades, the growing importance of commodities led to the stan-dardisation of the market and to individuals' dependency upon it.

However, Buck-Morss (1989) argues that the urban space such as shopping arcades (and later the development of department stores; see Case study 2.3) offered new opportunities for women, who had largely been marginalised in city life and whose realm was considered to be 'the home'. Arcades were places where women could mix and consume, and hence life in the modern city offered greater freedom and diversity through the development of modern capitalism. Women were encouraged into the city as consumers, purchasing goods and services for their homes and for themselves (see Chapter 5). Both as shop workers and as consumers, the social codes that restricted women's activities elsewhere were less obviously felt in the city. However, as Benjamin notes, the arcades represented a space in the city that was in a sense disconnected from stricter delineations of acceptable behaviour elsewhere in the city. Being associ-ated with increased freedom, arcades thus also become sites of sexual transaction and transgression. Benjamin noted, for example, that in the rooms above hosiers', haberdashers' and milliners' shops, where young female assistants finished goods and prepared them for display, sex was often for sale, in order to augment meagre wages (Benjamin 1999: 40). Similarly, Jane Rendell *et al.* (1999: 169) note that fictional and non-fictional representations of London's arcades often depicted them as sites of dangerous sexual transgression.

Despite these conflicts and tensions, Mona Domosh (1996) asserts the impor-tance of the relationship between the development of the modern city and

ideologies of consumer culture based on gender. In research on the retail landscape of the late nineteenth-century New York City, Domosh describes the
shopping streets that ranged from Fifth Avenue to Union and Madison Squares,
extending west to Sixth Avenue and east to Broadway. Domosh shows this to
be an area dominated by ornamental architecture and grand boulevards flanked
by restaurants, bars, small boutiques and large department stores – an urban area
that developed around consumption. As such, this new retail area was functionally different from its predecessors due to being solely oriented to retailing.
Importantly, the area was marked by being architecturally distinct, with elaborate
decorative detail and shop-window displays oriented towards middle- and upper-
class women.

With their grand architectural statements, these new urban spaces relocated the
private world of the home right to the heart of the city, and in shop windows the
display of commodities from bedroom sets to crystal, from corsets to evening
gowns, brought the private into the public realm. Retailers thus targeted women
as their customers, and to maintain appropriate gender roles these urban spaces
were feminised. The visibility of the 'home' in the city's public realm was thus
important in order to maintain the social and cultural values of New York's bourgeoisie and ensure that shopping in central urban areas was deemed acceptable
for women and children. These shopping districts had to appear to be cultural and
civic spaces and not completely tainted by commercialisation (Domosh 1996).
Importantly, then, at this time in the development of the modern city the presence
of women and the domestification of the city's streets was a way in which the
bourgeois class legitimised their wealth and power (this will be returned to in
Chapter 5).

## Case study 2.3 **The department store**

One of the most important moments in the development of consumer culture was
the advent of the department store in the middle of the nineteenth century. Richard
Sennett (1977) suggests that department stores grew because of changes in
the production system: the factory allowed more and more goods to be made
quickly, and thus more efficient outlets were required. Industrialisation encouraged the existence of vast emporia in which practically anything could be
purchased. Shopping became a quite different experience in department stores.
Prices were fixed, there was free entry, and anyone with money could shop.
There was no haggling, no obligation to purchase – and the phrase 'just looking'
came into being. The size and huge range of new stores meant that it was

continued

**Table 2.1** *The department store*

| Before the department store | After the department store |
|---|---|
| Purchase obligatory: 'just looking' impossible | Purchase optional: 'just looking' becomes possible |
| Ultra-specialisation: each shop sells only one type of good | Ultra-generalisation: each department store sells a vast range of goods |
| Retailing governed by guild system: restricted number of goods available in artisanal system | Retailing in the department store is a response to the availability of mass quantities of goods produced by the factory |
| No competition between guild members | Competition between department stores |
| No fixed prices: bargaining obligatory | Fixed prices: bargaining impossible |
| Need-centred goods: goods neither displayed nor advertised | Desire-centred: display and advertising of goods becomes vital to successful retailing |
| 'Shopping around' impossible | 'Shopping around' possible |
| Exchange or return impossible | Exchange or return possible |
| Production-centred | Consumption-centred |
| Shopping restricted to one's local area | Department stores attract shoppers from all over city and beyond |
| Personal characteristics of seller relatively unimportant | Personal characteristics of sales clerk must match 'cultivated' image of the store |
| Public space generally male | Creation of new female public space for both shoppers and workers |
| Could not provide cultural identity for new middle class | Cultural identity for new middle class can be bought off the shelf |

*Source*: Corrigan (1997: 61)

possible to wander about for hours, fascinated by the inventive displays in a palace of fantasies, seduced by the goods which promised all manner of pleasures through their mode of display (and shoplifting became a problem amongst middle-class women). There were new smells and sights, and you could try things on. All classes of people could enter and each could buy things (see Table 2.1).

However, in reality, department stores provided the material means for the middle class in particular to stake out their cultural identity. Of importance was the respectable presentation of service staff and surroundings, which included an abundance of marble, carpets and ornaments – the refined sophistication of galleries with a luxurious, almost aristocratic, ambience. Department stores sought to replicate the design and décor of theatres and art galleries with bright lighting, and grandly designed halls. It is through their construction in terms of their relative safety and organisation that department stores became an important female public space, a realm where commodities and gender became intertwined.

Source: Corrigan (1997)

As well as arcades and department stores there are also other consumption activities that are important to the development of the modern city and the new retail districts. David Frisby (2001), for example, in describing the dramatic expansion of metropolitan life in Berlin around 1930, notes the important proliferation of cafés and their street-side consumption cultures. He describes the development of a utopian vision for urban life in Berlin via the development of boulevards and plazas and 'world' city squares (such as Alexanderplatz, Potsdamerplatz and the Platz der Republik). However, underpinning the growth of Berlin was also the expansion of traffic and roads, and the rationalisation of the city to suit industrial production. Hand in hand with the development of such production-focused infrastructure was the emergence of specific areas where shops, bars and department stores captured flows of pedestrians and consumers. Frisby argues that it was in these shopping districts (which featured arcades and department stores) that an individual, rather than a collective, consumption economy can be seen to emerge. Such areas sought to attract into the inner city middle-class consumers, and, in order to visibly occupy the city's streets through new forms of sociability based upon consumption, the proliferation of street-side cafés was an important factor. Frisby shows that the practices of sitting, drinking, meeting with friends, family, colleagues and strangers, as well as engaging in people-watching, were an important function of this new urban café culture.

It is also important to stress that such bourgeois-led developments were played out against a backdrop of social and demographic change. Mass migration that serviced the factories' needs, and the activities of entrepreneurs from ethnic minorities, combined in different ways to bring about very different urban cultures (see Case study 2.4). However, despite reformist zeal and desire from the new middle classes for social and spatial distinction, during this particular period of modernity, spectacle and tolerance were common cultural denominators in cities. This engendered a heightened sensitivity to marginality as an urban norm in popular restaurants, cafés, department stores, hotels and shops that seemed visibly to encourage new urban cultures (Zukin 1995).

## Case study 2.4 **Migration and consumption**

Beyond the bourgeois-dominated new urban retail districts, a key feature of the modern city was mass migration and the mixing of different ethnic groups. In New York, for example, entrepreneurialism, shop ownership and the social and economic integration of migrant residents was a key feature of midtown and downtown areas. The presence of shoppers, peddlers, store owners, managers and clerks from all over Europe (and elsewhere in the world) led to the development of shopping streets that fostered new identities through interaction among, and fusions between, various ethnic traditions. The mix of different languages, cultural and religious practices, food and drink, music, and so on brought about a new urban diversity and consumer cultures underpinned by movements of commodities, people and ideas.

Glennie (1995: 184) argues that the migration of people into North America (with motives of greater political and religious freedom, economic improvement, consumer abundance and dreams of a better life) created 'consumption folkways' that led to new consumer identities. For example, during the nineteenth century, migration from Europe to America represented a movement from scarcity to abundance, and a key concern of migrants was to pursue social acceptance through consumption. For example, Glennie describes the distinctive response of European Jews to life in America. In Europe, Jewish religious tradition called for a selective use of luxuries (to be restricted to celebrating religious holidays). However, in the burgeoning consumer culture of modern American cities, 'luxury' commodities had become part of everyday life.

As such, for the wealthy amongst the Jewish community a new secular luxury emerged, where religious celebrations led to increased levels of consumption during religious holidays (augmenting the consumer comforts they enjoyed in their daily lives) through the taking of summer family holidays and the purchasing of expensive objects for domestic leisure, such as pianos, jewellery and elaborate furniture. Such consumption related not only to religious traditions but to newly unfolding ideas of the importance of family life and North American social refinement. For many other migrant groups, also, such mixing of European tradition and American life created new relationships between cultural identity, success, respectability and consumer goods.

What is important, then, is that the new consumption spaces that are associated with the modern city fostered a dramatic new urban culture based on acquisition, happiness and democratisation of desire, where money and bourgeois values dominated (Leach 1984). This, of course, had physical effects on the city through the architecture and design of both shops and entertainment sites. New materials and technologies were pioneered; plate glass, cast iron and steel construction, augmented with coloured electric lights, meant that 'stores embodied a sense of flux that many found disorientating' (Zukin 1998b: 623). The 'dramatic' design of these new places can be seen to have institutionalised the 'successful' identities of urban and commercial cultures. Increasingly these were being constructed through a sociability that was compatible with (and dependent on) the increased importance of a modern capitalist market economy. This was underpinned by the wealth created from the industrial revolution and the explosion in finance and banking. Analysis of the archetypes inhabiting these spaces has overwhelmingly been centred around the lifestyle of the much-written-about *flâneur*: an independent single man who strolled the streets, cafés, nightclubs and shops on the look-out for the new, exciting and unfamiliar. The flâneur is a stroller, whose purpose in window-shopping and displaying stylish purchases, including clothing, was nothing more than to see and be seen, a process of voyeurism and exhibitionism that characterised many major cities at the turn of the century. (Chapter 6 will return to the figure of the flâneur.)

Although there was no corresponding literary figure depicting the activities of the female *flâneuse*, the increase of the urban retail trade associated with these new shopping spaces meant that there were many new jobs for women, and places perceived as safe for women shoppers and children (Wolff 1985). Nevertheless, such opportunities led to special problems and a further range of archetypal identities. For example, sales clerks' salaries were often low, and women who had migrated to the city – faced with poverty, marriage or prostitution – often sought other available options to enhance their wages as actresses or nightclub performers (Zukin 1998a). Meanwhile, women customers were persuaded by merchants and fashion commentators to buy increasing numbers of things, and cities became more morally dangerous 'dream worlds for both men and women' (Williams 1982: 67). Although merchants greatly expanded access to urban public spheres and the publicity of shopping spaces, such spaces were undoubtedly founded on desire for money and material things (Leach 1984).

This period is marked as the birth of consumer culture through the proliferation of mass production and mass participation in consumption. While this was based on values and aims inherited from earlier periods, it represents a spreading of culture that was already well defined in other classes. However, the transformation of modernity itself into a commodity, of its experience and thrills into a ticketed spectacle, of its domination over nature into a proliferation of domestic comfort, of its knowledge into exotic costume, and of the commodity into the goal of modernity, was fully imposed at this time.

## Late modernity (1930–1975)

The 1920s, the culmination of around fifty years of the steady growth of consumer culture, was the first decade of full consumer affluence that promoted the powerful link between consumption and modernisation. From the 1920s, the world was to be modernised through consumption; consumption culture itself was dominated by the idea that everyday life could and should be modern. Aligned with this was a burgeoning advertising and marketing culture selling not just consumer goods but also consumerism itself as a shining path to modernity (Slater 1997: 13). Advertising incited people to modernise themselves, their homes and their means of transport.

The exemplary goods of this period were about the mechanisation of everyday life – electrification, durables, washing machines, vacuum cleaners and automobiles for that modern sense of movement into the future (see Ross 1996). This is the age of real estate, consumer credit and cars; modern appliances bought by modern methods, placed in modern households. Consumer culture from the 1920s to 1950s was about conformity – mass cultural dopes keeping up with the Joneses through slavish mass consumption of standardised mass-produced goods. As Chapter 1 showed, the consumer was born from the 'affluent worker', preoccupied with building up domestic capital and long-term job security (Goldthorpe et al. 1969). The goal of a stable family life and a preoccupation with the everyday needs of the household not only were an important source of continued demand for Fordist products, but ensured the co-operation of employees. The family became an important unit of consumption, and supporting and maintaining family life became a central driver of economic activity, underpinning business and organised labour. This completed a colonisation of everyday life by corporations and consumption norms which rendered employees as status-driven and conformist.

However, just as consumer capitalist mass-production and consumption cultures were established, the Wall Street Crash in 1929 and the subsequent 1930s depression, and broader social upheaval such as the civil rights movement, created unrest and change in the city. This led to further dramatic changes to the city and consumption cultures due to processes of suburbanisation and 'white flight' from the city. The Wall Street Crash was due to an inherent weakness in the capitalist accumulation of the period, in particular to the overproduction of goods and problems relating to money and the stock market. First, then (and simply), companies were producing too many goods; American goods could not be sold abroad because other countries had put tariffs (taxes) on them to make them more expensive, and when demand for goods began to fall, workers' wages were cut and some workers became unemployed, which meant that they could no longer afford to buy the new consumer goods.

Secondly, people were allowed to borrow too much money and could not afford to pay it back. People had taken out loans or invested their savings in the stock market while there were too few controls on the buying and selling of shares. Advertising and hire purchase agreements were not controlled, and this encouraged people to spend more. Banks did not have enough money in reserve to help businesses that were in trouble. Having lent too much money the banks were facing difficulties because people could not afford to repay their loans. In the context of urban civil unrest, poverty, unemployment, urban decay and crime, the relative proximity of middle-class districts to working-class ghettos and the previous liminal mixing in spaces and places of consumption was ended, as those with the capacity to leave began to move to the growing suburbs. The remainder of this section will look at both the inner city and the affluent suburbs.

## The inner city

Le Corbusier's vision of the modernist city was about replacing the crumbling disorder of existing urban environments and obliterating any link to their urban pasts. Stevenson (2003) argues that while the high-point for this modernist architectural theory was the period between the two world wars, the high-point for the practice of modernist architecture came initially in Europe after the Second World War reconstruction, and later in the rest of the world. After the

Second World War Le Corbusier's ideas shaped urban redevelopment and reconstruction initiatives around the world, and particularly high-rise estates built for low earners in many countries. The concrete modernist vision of urbanity legitimised massive slum clearance and the creation of large municipal housing estates on greenfield sites.

However, by the 1970s Le Corbusier's dream had unravelled. Housing estates designed and constructed according to the principles of modernist architecture were acknowledged as social failures, and some had even been demolished. Ideas of community, participation and local identity, which were anathema to the utopian dream of standardised collective consumption, had proved to be very important in maintaining social coherence and values, and without the sense of community the corralling of the poorest into high-rise high-density estates had disastrous consequences. The high concentrations of social exclusion, crime, violence and unemployment in these areas ultimately led to social unrest and polarisation around the fault lines of class and ethnicity.

Stevenson argues that one specific moment signalled the end of the modernist utopian vision of a scientifically planned and rationally ordered city. In St Louis, Missouri, on 15 July 1972 at 3.32 p.m., the by-then infamous Pruitt-Igoe housing project was dynamited. Previously the high-rise buildings had been vandalised, mutilated and defaced by their black inhabitants. The surrounding parklands and recreational space was plagued by violence and crime. Although millions of dollars had been spent in order to try to maintain the housing by fixing broken elevators, repairing smashed windows and repainting, the high concentration of poor disenfranchised people, often on low wages or unemployed and living in squalid conditions, led to social unrest and ensured the economic and cultural isolation of areas that became known as ghettos (ghettos will be returned to in Chapter 3). Despite the important symbolic destruction of the Le Corbusian Pruitt-Igoe, large numbers of similar projects remained in cities throughout the world. Whether in inner-city areas or the large municipal housing estates that are found on the edges of our cities, the visions of modern urbanity that are represented by these planned communities continue to impact on the contemporary city.

## The suburbs

The growth of the suburbs represents a fundamental social shift of the consumer society that radically altered the nature of the great metropolis to a suburban, decentred, automotive-bound city (especially in the United States). The archetypal consumption space here became the shopping centre: a 'multi-purpose greenfield development surrounded by fairly homogeneous residential communities' (Zukin 1998a: 828). As with the early twentieth-century development of city-centre shopping districts, suburban centres relied on innovations in transport, building and display. As a consequence, the suburbs soon drained investment from city centres, and drew in both urban and suburban shoppers who would previously have shopped in the city centre. In essence, this period was characterised by a suburban synthesis of mass consumption and family-oriented lifestyles, which provided a cultural context for even more rapid development, leading after 1945 to socially and visually homogeneous shopping centres and malls (Zukin 1998a).

While suburban malls simulated the city's historic consumption spaces and attracted greater investment and popular patronage than the highly heterogeneous consumption spaces of the city, malls can also be seen to have 'reversed modern urban consumption trends' (Zukin 1998a: 829). Instead of dominant places of consumption being central to mass transport systems, malls were primarily accessible by *private* transport. In the 1950s, malls were eventually roofed and built with plate glass, electric lights and air conditioning – the same

**Figure 2.4  A suburban estate, Newcastle upon Tyne, UK. (Courtesy of Simmons Aerofilms Ltd)**

materials as the great department stores – but made more comfortable, with landscaped trees, plants and waterfalls. Such traditional urban characteristics can be seen as privatised 'quasi-streets' with street furniture and food courts in order to detain shoppers longer. However, central to the success of suburban shopping centres was a cultural package of family privacy, employment and the primacy of cars. The sociability fostered by shopping malls often depended on groups rather than individuals. Non-working women met, elderly people benefited from climate control and guards, and 'a more fluid network' of friends and ages demonstrated malls' usefulness as public spaces (Zukin 1998a: 830). Similarly, the patrols and curfews on teenagers enforced by security guards presented a safe, comfortable environment for respectable shoppers.

During the 1970s and 1980s, shopping malls developed two distinct spatial forms, according to Zukin (1998a). First, increasing competition from discount chains led to wider choice and lower prices in both freestanding superstores and larger hypermarkets. Secondly, fairground rides and amusements and multi-screen or multiplex cinemas were built around the malls. Older malls were refurbished with enlarged food courts and more entertainment, in essence becoming more socially heterogeneous. New groups became attracted to these consumption spaces, and they became more racially and ethnically diverse, often drawing people from the city in search of a high-quality safe and clean shopping environment. Far from remaining local consumption spaces for a homogeneous residential community, malls took on new forms, becoming regionally significant and specialist outlets, filled with suburbanites, urbanites and tourists out for a good time (Zukin 1998a). Outside North America, the proliferation of these spectacular malls was a little slower and on a less grand scale.

Moreover, the suburbs offered a calmer, safer, more prosperous way of life – in direct contrast to the unstable city. Newly built houses in the suburbs offered the perfect shell to fill with consumer durables. The suburbs offered a further separation between domestic life and work. White middle-class commuters could concentrate on taking care of their property, and their families could avoid the public and socially mixed street life and reduce their fear of strangers. The suburbs were dominated by roads and cars with distinct private space. This eliminated or marginalised the tensions of urban life, separating out different geographies and histories, and de-intensifying social interaction – what has been described as an anti-urbanism, in its opposition to the characteristics which had defined 'city-ness'.

The suburbs were built for domestic life, for making the most of leisure time, the weekend and the family. The straight, regular avenues ensured houses had space on either side and were separated by large fences. The suburbs were places of private family lives, but this often led to isolation; public transport was oriented to getting to and from the city, and not within suburban areas. This meant that social relations were built around networks and clubs. The famous Tupperware parties, for example, sought to create economic and social networks for isolated middle-class suburban women (Clarke 1997). It was in the suburban houses, streets and malls that white middle-class respectability was protected and maintained.

## Concluding remarks

It is in the modern city that consumer culture and the core political, economic, social and cultural institutions, organisations, infrastructure, practices and processes of modern life were inextricably bound together. The strengthening and formalisation of local municipal government power, and the growth of a modern market economy become aligned with social and cultural identities, spaces and places. It is this strengthening of bourgeois political control and Fordist economic growth and the development of culture- and consumption-oriented activities that dramatically impacted on the development of the modern city-scape.

It is in the modern city that consumption became bound up with the very idea of what the modern world was about; in particular, how individuals experience and view selfhood, identity, belonging and distinction from other social groups and the spaces and places they inhabit. Consumer culture therefore did not simply materialise from the industrial and intellectual success of modern thought – but rather 'the consumer' and the experience of consumption became integral to everyday modern life and the structure and organisation of the city itself.

Learning outcomes

- To be able to explain the relationship between consumption and the development of the modern city
- To be able to identify the new archetypal urban spaces and places associated with consumption cultures
- To understand the relationship between production and consumption and the social control of the city
- To be able to describe the relationship between new archetypal identities, lifestyles and forms of sociability, consumption and the modern city

## Further reading

Peter Corrigan (1997) *The Sociology of Consumption*, London: Sage. This book provides a clear and lucid review of theoretical and case-study-based approaches to studying the social relations that surround consumption.

Don Slater (1997) *Consumer Culture and Modernity*, Cambridge: Polity Press. A useful and accessible review of wide-ranging conceptualisations of the relationship between modern life and consumer culture.

Alan Tomlinson (ed.) (1990) *Consumption, Identity and Style: Marketing, Meaning and the Packaging of Pleasure*, London: Routledge. A review of the relationship between consumption and identity formation and social selfhood that draws on theories of broader urban change.

David Frisby (2001) *Cityscapes of Modernity*, Cambridge: Polity Press. A book that marries an understanding of the physical development of the modern city and the presence of its archetypal protagonists. Draws on theoretical debate, popular culture and contemporary accounts of the development of the modern city.

# 3 Consumption and the postmodern city

**Learning objectives**

- To understand the relationship between consumption and the development of the postmodern city
- To be familiar with archetypal postmodern identities, lifestyles and forms of sociability
- To describe the new urban spaces and places associated with postmodern consumption cultures
- To offer insights into the role played by consumption in the increasing social and spatial polarisation of the city

This chapter will show that over the past twenty years the nature of everyday urban life has been profoundly affected by the global reconstruction of economic, political, social and cultural processes. Related to this profound change has been the decline of the heavy and manufacturing industries that dominated the modern city and an increase in the importance of post-industrial service industries such as financial services, banking, advertising, marketing, public relations and the retail sector. This has been coupled with social and demographic forces that saw the simultaneous increase in mass unemployment, and the rise of a 'new petite bourgeoisie' (Giddens 1973). These processes have been reflected in new spatial and social formations, and academic attention has been concerned to map their impact upon the changing city. Theorists have described these changes as being bound up with a movement to late or advanced capitalism – from a modern to a postmodern epoch.

One of the central tenets of this conceptualisation of a 'new' global spatial and symbolic urban economy and hierarchy is that the city, which historically was politically, economically, socially and spatially organised around production, is now said to be underpinned by consumption. Thus, as cities attempt to combat long-term decline and develop a sustainable post-industrial economy with business, financial and professional sectors, it is argued that there has been a complementary shift to emphasise the service sector. Alongside this shift, there has also been an increasing growth in the importance of cultural industries, and the production of consumption has come to the fore (Barke and Harrop 1994). Hence, the economy of the postmodern city is less based on production and consumption of goods, more on the production and consumption of culture. The significance of culture is linked to the rise of a symbolic economy concerned with making and distributing images (Scott 2000). In the postmodern city, the projection of image lies at the heart of the attractiveness of style in the city. In contrast to modern cities, where function shaped appearance and where products and buildings were mass-produced and generally standardised, in the postmodern city style, design and appearance rule.

## Conceptualising the postmodern city

These new socio-spatial urban configurations have been presented as being very different from their predecessors (Zukin 1982; Knox 1987; Soja 1989; Harvey 1989b; Davis 1990). This 'new' city is visibly more spectacular: revitalised city centres and agglomerated business and financial districts feature gleaming high-rise office blocks, waterfront developments, flagship buildings such as concert halls and museums and 'urban villages' – such as London's Canary Wharf, Barcelona's Olympic Marina, Paris's La Défense, Vancouver's Pacific Plaza, New York's Battery Park (see Figure 3.1) and Sydney's harbour. Moreover, these economic and symbolic city spaces and places are surrounded by others characteristic of postmodern urbanity. These include high-technology business clusters, out-of-town mega-malls, elite 'gated' residential neighbourhoods, ghettos, and 'edge cities': master-planned 'suburban' developments with town centres, public squares, police and fire stations, and other activities previously only found in cities – commercial functions that become decentralised due to high land costs and social and physical decay associated with inner-city areas. As Figure 3.2 shows, the postmodern city is decentralised and de-differentiated: the specialised zoning characteristic of the modern city has declined, as activities that used to be differentiated by their concentration in particular parts of the city have been dispersed. Table 3.1 compares the main features of the modern in contrast to the postmodern city.

With the decline in the traditional industries that underpinned the economic potency, political organisations and social and cultural make-up of modern urban

Figure 3.1 *Manhattan seen from Battery Park. (Courtesy of Malcolm Miles)*
*Source: Miles (1997)*

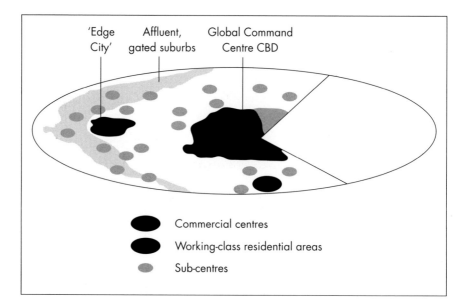

**Figure 3.2** *The post-industrial 'global' metropolis. (Courtesy of Tim Hall)*
*Source: Hall (1998)*

life during the 1980s, North American and European cities were hit with recession that led to unemployment and physical decline. This 'urban crisis' was met with 'a new urban politics', which saw city authorities (previously focused on local provision of welfare, services provision and collective consumption) being forced to adopt a more outwardly oriented stance designed to foster local growth and economic development (Harvey 1989b). Policies described as 'neo-liberal', including risk taking, inventiveness, and the driving necessities of promotion and profit, engendered an entrepreneurial outlook by local government agencies, and led to an increase in the privatisation of service provision in an attempt to respond to economic, social and cultural change.

**Table 3.1** *The modern–postmodern city: a summary of characteristics*

| **Urban structure** | |
|---|---|
| *Modern* | *Postmodern* |
| Homogeneous functional zoning | Chaotic multinodal structure |
| Dominant commercial centre | Highly spectacular centres |
| Steady decline in land values away | Large 'seas' of poverty |
|    from centre | Hi-tech corridors |
| | Post-suburban developments |

## Architecture, landscape

*Modern*
Functional architecture
Mass production of styles

*Postmodern*
Eclectic 'collage' of styles
Spectacular
Playful
Ironic
Use of heritage
Produced for specialist markets

## Urban government

*Modern*
Managerial – redistribution of resources
  for social purposes
Public provision of essential services

*Postmodern*
Entrepreneurial – use of resources to lure
  mobile international capital and
  investment
Public and private sectors working in
  partnership
Market provision of services

## Economy

*Modern*
Industrial
Mass production
Economies of scale
Production-based

*Postmodern*
Service-based sector
Flexible production aimed at niche
  markets
Economies of scope
Globalised
Telecommunications
Finance
Consumption-oriented

## Planning

*Modern*
Cities planned in totality
Space shaped by social ends

*Postmodern*
Spatial 'fragments' designed for
  aesthetic rather than social ends

## Culture/society

*Modern*
Class divisions
Large degree of internal homogeneity
  within class groups

*Postmodern*
Highly fragmented
Lifestyle divisions
High degree of social polarisation
Groups distinguished by their
  consumption patterns

Source: Hall (1998: 89)

As a result, increased competition has arisen around the efforts of cities to create new images in order to attract speculators, businesses and consumers. Cities have embarked on advertising campaigns and marketing strategies that have been augmented by speculative developments and partnerships with private capital (see Chapter 6). If cities were to successfully compete in a new postmodern urban hierarchy, local authorities became aware that they would be forced to initiate redevelopments that were often speculative in nature, high profile, and had symbolic significance in attempting to represent progressive socio-cultural and economic trends – to develop new economic, social and cultural conditions which were attractive to post-industrial employers, investors and tourists.

In line with this, there was also a raft of other physical and symbolic attempts to improve the urban environment, which Landry (1995) considers as initiatives to promote a 'creative city'. These encompass aesthetic improvements of 'soft infrastructure', ranging from the building of squares and fountains to the greening of streets, the provision of benches and improved public spaces, and the estab-lishment of late-night shopping, 'happy hours', cultural events and festivals such as music, literary or street theatre – all designed to make the city more 'live-able'. With buildings and facilities such as theatres, art galleries, convention and exhibition centres, as well as a supporting cast of café bars, restaurants, fashion boutiques, delicatessens and other cultural facilities, the buzz of 'creativity, innovation and entreprenurialism' – which is thought to be associated with these activities – is seen as crucial to contributing to the competitiveness of cities.

It is clear, then, that such restructuring of cities – particularly those that have most successfully moved from a focus on manufacturing production to a service- and consumption-based economy – is founded on particular social and spatial forms of urban life. Central to such strategies of urban regeneration are processes of possession and visibility relating to the production of city images and cultural representations, and to the nature and use of physical spaces/places of the city. In other words, this is a construction and imagining of the political, economic and symbolic centrality (and marginality) of people and places which can be summarised as 'an idealised white middle-class hegemonic notion of urbanity, often reclaiming or gentrifying marginal spaces, leading to the displacement and further exclusion of marginal populations, the re-definition of collective memory or (more rarely) commodification of minority ways of life' (Hall and Hubbard 1998: 110). In essence, as with the modern city, the postmodern city is organised around social relations where some are empowered and others disempowered.

The fragmented and decentralised city is thus also characterised by increased segregation of different populations as cities became more divided along the lines of inequality and ethnic diversity. While some urban spaces are developed around the consumption practices of the 'new petite bourgeoisie', in other parts of the city (such as inner-city or suburban residential areas) there is an ever-widening gulf between the 'haves' and the 'have-nots'. For example, the declining industrial city with unemployment, fiscal problems from the erosion of the tax base, and poor households in poor areas is characterised as being plagued with housing abandonment, arson, vandalism and sometimes riots. Moreover, new kinds of flexible deregulation of the labour market ensure that it is only low-paid, insecure and low-status jobs that are generated by the demands of gentrifiers, conference delegates and other affluent consumers.

Case study 3.1 explains how these new spatial and social relations are under-pinned by the shift from Fordism to a more flexible mode of capitalist production and accumulation. It is through this movement to postFordist accumulation that these economic conditions and differential employment opportunities develop, and hence lead to the increasing social and spatial polarisation in the postmod-ern city, where landscapes of consumption and devastation exist in intimate relationship to one another (Zukin 1982; Harvey 1985; Davis 1990).

## Case study 3.1 **From Fordism to postFordism**

Up until the 1970s, consumer society was thriving through the system of Fordist mass production and consumption that had overcome economic crises and trade cycles such as the 1930s depression. However, once again in the early 1970s it became clear that the system had reached its internal and external limits. Costs and timescales of Fordist production and the logic of 'high output, low cost' were pushed to their limits, and goods were being sold with decreasing margins due to an increasingly saturated consumer market and faster turnover of fashions, tastes and trends – which is itself the result of successful Fordist advertising and marketing.

The response to this crisis in Fordist capitalism was to move to a more flexibile mode of production – postFordist flexible accumulation. In place of dedicated plant churning out high volumes of standardised goods in order to achieve economies of scale, the aim now was to have flexible labour and plant that could cost-effectively produce smaller batches of more customised goods. The time it took for goods to be designed, produced, advertised and sold had to be greatly reduced. New technologies such as robots and computerisation allowed flexi-bilisation to extend beyond the production line to include flows of information. PostFordism is associated with rapid and interconnected flows of information from

continued

points of sale through to material suppliers. This is known as 'just-in-time' management and relies on such flows to ensure precise or daily (even hourly) breakdowns of sales of every item in any product range. This is important in order to ensure that only what is needed is produced. This change in production was fuelled by the globalisation of consumer goods markets, faster product lifecycles as consumers desire more and more fashionable innovation, and thus far greater product/market segmentation and differentiation. Fordism (characterised by mass production, consumption, modernisation, uniformity and standardisation) thus gives way to postFordist production and consumption cultures.

Lash and Urry (1994) show that postFordism is demand-driven, generates a variety of product types, and is characterised by individualisation of taste and lifestyle – pursued through the consumption of 'designer' labels and brands, where the goods and services 'are not material objects but *signs*' (Lash and Urry 1994: 4). This represents an 'aestheticisation' of everyday life that is argued to be founded on, in simple terms, the demands of a burgeoning middle class, and particularly the new middle classes of managerial, professional and service employees, who sought to distance themselves from the working and lower middle classes who were widely consuming mass-produced goods and services (see also Du Gay 1997; Jackson and Thrift 1995; Molotch 1996). PostFordist markets are not aggregated or undifferentiated, but based on 'lifestyles', 'niche markets', 'target consumer groups' – not defined by broad social demographic structures such as class, gender and ethnicity.

The development of the modern city was in large part fuelled by the new commercial and trading classes associated with industrialisation and later the new white-collar and middle management – a class fraction created by the increasing bureaucratic complexity of the Fordist firm, and constituting initially the market for Fordist mass production. The movement to a postmodern era and the associated transition to postFordism are thus related to the emergence of a new middle class, predisposed to have an affinity with postmodern culture and consumption.

These are new bourgeoisie 'go-getting' executives working in finance, design, marketing and non-material sectors of production as well as new cultural inter-mediaries – workers in presentation and representation, symbolic goods and services, media and advertising, public relations and marketing, etc. – these identify and produce exemplars of postmodern culture; they are the communicators and sign producers who have displaced the commodity producers of organised capitalism (Lash and Urry 1987). Featherstone (1991) argues that, despite their economic success, the new petite bourgeoisie lacked the confidence of more established middle-class groups. The new bourgeoisie is characterised by a constant self-monitoring and self-consciousness. Featherstone argues that the new bourgeoisie attempted to compensate for this lack of confidence by investing in a self-constructed expertise in style and lifestyle. For Featherstone (1991) this makes the new bourgeoisie natural consumers of music, holidays and fashion. Such cultural activities both represent and legitimise the status and privilege offered by middle-class occupations.

The rest of this chapter will focus on the development of the postmodern city characterised via the relationship between consumption and archetypal identities, lifestyles, forms of sociability, and urban spaces and places noted above. However, what is important to grasp from Case study 3.1 is that underpinning the relationship between consumption and the development of the postmodern city is the idea that we have reached a new stage of consumerism. This 'new', accelerated consumer culture based on *individualised consumption* and its symbolic and aesthetic relation to the emergence of the new middle class on the one hand, and mass unemployment, low wages and insecure employment on the other, has had a profound effect on every aspect of urban life (Baudrillard 1975; Zukin 1982; Bourdieu 1984; Harvey 1989b; Featherstone 1991; Maffesoli 1992; Lash and Urry 1994). Before looking in more detail at the changes associated with the development of the postmodern city, it is thus important to pause to look at the theorists and theories that help to explain these changes in consumer capitalism and their impact on the city.

## Conceptualising the relationship between consumption and postmodernity

There has been some difficulty in reaching a clear consensus about what it means to talk about postmodern times. Some see postmodernism as a response to a perceived collapse of modernity as a cultural project, a failure of modernity to live up to its goals of progress and achievement and improvement of people's daily lives (Jameson 1991). Others consider it a new turn of modernity itself, a redirection or re-orientation of those goals operationalised through political, economic, social, cultural and spatial change (Harvey 1989a). For some, postmodernism is simply bound to the economic restructuring associated with the move from an industrial to a post-industrial and Fordist to a postFordist economy (Lash and Urry 1987; Harvey 1989a). For others, the development of postmodernism is solely a cultural dynamic, and is not a response to economic restructuring. As such, no fundamental break is premised, and postmodern cultural forms exist alongside those associated with modernism itself (Berman 1992; Pred 1996).

However, while there has been much debate over the past decade, a consensus appears to be reached that postmodern times exist differentially in different places, emerging at different times in a variety of positions between these polarities (Lash 1990; Featherstone 1991; Giddens 1991; Lash and Urry 1994). Importantly, research relating to urban change, identities, lifestyles and forms of sociability has been central to reconciling some of the differing theoretical debates about who, where and when might be described as postmodern. This section reviews these arguments.

While definitions of postmodernism vary, and the term means different things to different people, one certain thing is that there is widespread acceptance that consumption plays an important role in the emergence of postmodern culture. Insights from writers such as Simmel, Veblen and Bourdieu regarding the formative role that consumption plays in social classification through consumption offered a valuable understanding of the development of modernity. However, although postmodern theory suggests that such social classification has become redundant, writers agree that consumption is a key element in the move from a world dominated by such modern structures to a postmodern period. This consensus is based on the recognition that consumption has become increasingly more significant for its symbolic qualities (or sign value) than for its actual functional role (use value).

It is in these terms that consumption is argued to be a new cornerstone of a new cultural code in a world where social classifications based on 'fixed' categories such as class, gender and ethnicity have collapsed – where everything can be a consumer item, including meaning, truth, knowledge, and our identities and selves. For instance, Mike Featherstone in his book *Consumer Culture and Postmodernism* (1991) argues that in a postmodern world everything is possible – we can be whoever we want, as long as we are able to consume. This is enabled by the 'aestheticisation' of everyday life, where 'good style', taste and design are part of our everyday lives, and 'cultural intermediaries' such as advertising, marketing and the media present a world where we are led to believe we can construct our lives, our very selves and our identities, and reflect this to the world through consumption.

This contrasts to the modern social system that involved the constitution of different social realms and groups (each with autonomous development and progress that constituted the differences and distinctions between them) through social or conceptual hierarchies. In postmodern cultures these realms and distinctions blur or implode so that distinctions and boundaries that make up our understanding of what constitutes categories such as class, gender and ethnicity are much more blurred and difficult to define. Lash (1999) argues that central to this process is the dematerialisation of objects and the 'triumph of signs' that leads to the instability, malleability and fluidity of culture. What Lash means is that there is a breakdown or implosion of the difference between representation and reality, sign and material good, culture and economy.

It is thus a central theme of postmodernity that a breakdown, instability or transgression takes place between boundaries and distinctions. It is in the world of consumer culture and on our cities' streets that this eclecticism and blurring is seen in everyday life. For example, whereas in the modern city particular food, clothing, musical tastes, ways of speaking and other leisure activities were more easily attributable to a certain social group, there is now much more hybridity and juxtaposition. Central to this is a transformation of reality into images and fragmentation where aesthetic experience becomes the master narrative.

However, while such processes are determined and driven by particular middle-class sensibilities, they have come to dominate production of culture and spaces and places in the city. Consumption is thus a central part of a social process that goes beyond the direct control of specific groups. Consumption provides everyone with a sense of control, and some semblance of authority over our lives. It is this process of commodification that underpins the move to postmodern times, where image, product, style and design have taken over from meta-narratives of conferred meaning that had previously bound together social groups. In these terms, then, the idea of postmodern culture is the domination of information, media and signs, the desegregation of social structures into lifestyles, and the general priority of consumption over production in everyday life.

## Consumption and the 'end of the social'

Jean Baudrillard (1988), perhaps the most high-profile writer on the topic, takes this theorisation of postmodernity to what he sees as a logical conclusion. Baudrillard describes consumption as an apprenticeship for social indoctrination that is based on a specific mode of socialisation. He argues that new social relations are associated with new productive forces and the emergence of a new economic system. Baudrillard argues that human needs are never satisfied and never fulfilled. It is in this context that Baudrillard argues that consumer objects increasingly take on the value of being a sign. Elaborating on this assertion, Baudrillard discusses the example of a washing machine, which he says is no longer tied to a function or defined need, but responds to 'a logic of desire' (1988: 44).

Consumer objects exist in a world of fashion where goods signify the potential to fulfil human desire, but can never actually do that. Baudrillard argues that what emerges from this is a constant fluidity of differential desires and meanings. On this basis, Baudrillard suggests that consumption has nothing to do with pleasure. Pleasure, he argues, is thus institutionalised as a duty on the part of the consumer (see Figure 1.1). In essence, Baudrillard's argument is that signifiers of economic value (i.e. currency) have become entirely divorced from any necessary relationship with the signified's use value. What is therefore created is an 'aesthetic' hallucination of reality (1993). It is in this way that consumption represents a means of expressing dream-like representations. Baudrillard suggests that the mass media have an important role in these representations and that consumption is part of that superficiality where that hallucination is perpetuated.

As such, for Baudrillard there is no reality left, only a code to be interpreted – appearance without depth – and postmodernism is not simply a culture of the sign, rather it is a culture of what he describes as 'simulacrum'. For Baudrillard, the simulacrum is an identical copy without an original – in fact, the very distinction between the original and the copy has itself been destroyed. He calls this process 'simulation' and, further, he suggests that the distinction between simulation and the 'real' implodes, and the 'real' and the imaginary continually collapse into each other, thus producing the hyper-real. The result is that reality and simulation are experienced as without difference; simulations can often be experienced as more real than the real – even better than the real thing.

While he suggests that evidence for hyper-reality is everywhere, Baudrillard describes Disneyland as perhaps the best example of hyper-realism. He argues that the success of Disneyland is not due to its ability to allow Americans a fantasy escape from reality, but because it allows an unacknowledged concentrated experience of the 'real' America. In these terms, Disneyland is there to conceal the fact that it is the 'real' country. Disneyland is presented as imaginary to make us believe that the rest is real, when in fact Los Angeles or Las Vegas (see Case study 3.2) and the America surrounding it are no longer real, but of the order of hyper-real and simulation. It is no longer a question of false representation of reality, but of concealing that the real is no longer real. Baudrillard explains this in terms of Disneyland's social function – a postmodern collapse of certainty, a dissolution of meta-narratives such as truth, god, nature and the law, which have lost their authority as centres of authenticity and truth and been replaced by hyper-real simulacra. As Case study 3.2 shows, representation does not stand at one to remove reality, but is reality.

## Case study 3.2 **Viva Las Vegas**

Las Vegas, like many other American towns, grew along with the railroad. In the 1930s a series of events conspired to accelerate its expansion (such as the Los Angeles authorities seeking to clean up the city in 1938 to get rid of prostitution and gambling, which moved to the desert), and from that time Las Vegas became synonymous with gambling and entertainment culture, show business and the likes of Elvis, Tom Jones, the Rat Pack, boxing, quickie marriages, risqué floor shows and gangsters.

The past twenty years has seen a boom in the fortunes of the city. Boasting its one-hundred-thousandth hotel room, a figure unmatched anywhere in the world, the city has had a whole host of new developments. Some of the 'simulacra' included in this development include the Bellagio, a $1.4bn version of a north Italian resort (with lake), a fifty-storey replica of the Eiffel Tower and other Parisian landmarks such as the Champs Élysées, and a $1bn tropically themed casino. Las Vegas

**Figure 3.3** *The Luxor, Las Vegas. (Courtesy of Taylor & Francis Books Inc.)*
*Source: Rothman (2002)*

constitutes a fantasy land and has recently fully emerged as a Disney-style theme park. For example, Buccaneer Bay, a $30m artificial lake outside the Treasure Island casino-hotel, hosts a full-scale battle between Spanish and British frigates; nearby, at the Caesar's Palace casino, The Forum is a Roman-themed shopping street crowned by a vaulted ceiling painted to look like natural sky, and its appearance is changed by computerised lights. Casino architecture is based on fantasy, such as the Egyptian pyramid of the Luxor, completed in 1993.

Postmodern theory would suggest that it is no longer possible to identify the boundaries between these representations and fantasy. Much of the writing on postmodern urbanity identifies Las Vegas as the epitome of hyper-reality, a space where representation itself has become more real than the reality it ostensibly depicts. In *America* (1988) Baudrillard argues that places such as Las Vegas and Disneyland hide the real America and produce a hyper-real order of simulation. He argues, however, that it is the spaces and places of Las Vegas and Disneyland that have now become the authentic America – it is now the clean-cut, themed and policed environments of Disney that dominate urban experience. The streets, casinos and hotels of Las Vegas provide a rich and varied case study to develop Baudrillard's ideas, and can perhaps be seen as a perfect exemplar of a postmodern fantasy city.

As such, relating postmodern experience to social structures or to the structural positions of social groups based around, for instance, class, gender, ethnicity and sexuality is said to be defunct. Postmodern experience serves to explain the disappearance of social groups based on such categorisation, replaced by a consumer culture where we can buy and wear our identities as we wish. In these terms, postmodern experience is said to be total because it has swallowed up 'the social' which would otherwise differentiate it.

## Are we all postmodern?

While Baudrillard's arguments are undoubtedly persuasive and take debates around postmodern culture to their logical conclusion, what cannot be ignored is that the world is divided by wealth and power. Thus, while depictions of postmodern urbanity such as Las Vegas suggest that experience is solely based on access to consumer goods, and more generally the ability to construct one's life along the model of the consumerist life and inhabit particular spaces and places within the city, then money and culture do very directly restrict access to postmodern culture and postmodern urbanity. The question must be asked: how pervasive is the argument that social differentiation based on modernist classifications has been replaced by a new order based firmly on consumerism? If this is true, then is everywhere, and everyone, postmodern?

Critics of Baudrillard would argue that a more fruitful question would be to ask: who is postmodern, when and under what conditions, and where? For example, in line with a more cautious depiction of postmodernism, in *Postmodernism or the Cultural Logic of Late Capitalism*, Frederic Jameson (1991) argues that such postmodern approaches fail to look at experience and the role consumption plays for individuals in (and individual experiences of) capitalism. Baudrillard's depiction of postmodernity is thus limited by a dependence on abstract theories that don't actually contribute to an understanding of everyday life and how consumption affects people's experiences of postmodernity. Frederic Jameson suggests that late capitalism amounts to a 'purer' form of capitalism than previously experienced, and that there is a fundamental relationship between the positioning of postmodernism in the economic system and its impact on the sphere of culture, including contemporary society. He argues that aesthetic production is integrated with consumption in order to produce fresh waves of newness, with consumer goods and services being produced at ever-increasing rates, with aesthetic innovation and experimentation as key ingredients.

The key to Jameson's analysis is that it is misleading to suggest that contemporary society is about nothing more than cultural difference and fragmentation. He is a Marxist and argues that the production of culture has been subsumed into commodity production, and in this sense postmodernism equates with late multinational consumer culture. This suggests an increasingly close relationship between so-called postmodern forms of consumption and unequal social and spatial relations that pervade structures of consumer capitalism and hence spaces in the city (see Case study 3.3).

## Case study 3.3 **Consumption: collective and individual**

Theorists wishing to argue for a shift from modern conditions to those of postmodernity suggest that a departure from collective consumption takes place. This suggests that 'aestheticisation' of everyday life (Featherstone 1991) produces a new kind of 'tribalism', where membership is marked by paraphernalia (clothes, taste, music, fashion, etc.) and a spontaneous structuration in localities (Maffesoli 1992).

This in effect implies a re-inventing of city life, where modern forms of socialisation and collectivism have been replaced by competitive, *individualistic* forms of market

continued

consumption. This 'tribalism' is identifiable in the spaces created by urban regeneration projects globally. Critics have sought to consider the extent to which urban populations are now divided by lifestyle rather than social class, gender, race, ethnicity, sexuality, and so on. As such, spaces of consumption (whether pedestrianised city streets or out-of-town shopping centres) have been transformed into playgrounds for new forms of 'collective' activity, wherein individuals can reaffirm their existence through consumption. While this is said to create a collective sense of belonging (based not on class or kinship but on consumption), certain groups without cultural or economic capital are excluded (Bourdieu 1984). The most prominent groups associated with this growth of service industries relate to the demographic influence of the baby-boom generation, which includes the 'new middle class' demanding distinctive high-quality goods and services (Zukin 1998a: 825), and an associated increased visibility of previously marginalised groups – such as lesbian and gay, ethnic and youth groups (Zukin 1995; Mort 1996).

However, theorists have undoubtedly masked the differentiated situations within many cities (and places/spaces within cities). In order to establish a framework for theorising new urban lifestyles, their relationship to/basis in both traditional 'modern' urban values or categories such as class, gender, ethnicity and sexuality, and their establishment in specific systems of values and social space, have been ignored.

Bauman (1987, 1998) takes up this point and distinguishes between the 'seduced' and the 'repressed' in consumer culture. In essence, Bauman is making a distinction between those who can and those who are less able to enter consumer culture. Bauman suggests that consumer choice represents the foundation of a new concept of freedom in contemporary society, and that freedom of the individual is constituted in individuals' role as consumers. Bauman argues that while contemporary consumption has opened the possibility of choice to increasing numbers of people to act as 'free' individuals, massive oppressions are also generated. Those who are economically excluded from making such choices become disenfranchised and oppressed. Bauman thus describes the 'seduced' as those members of society for whom consumption becomes a major arena of liberation. On the other hand, the 'repressed' are those who simply do not have access to the necessary resources to become involved (to the extent that they would like to) in all that the consumer society has to offer. This section of the population therefore becomes dependent upon the welfare and other state support services and institutions. Bauman concludes that a key feature of consumerism and the contemporary consumer society is division, and not only does the consumer society not offer the same opportunities to all, but it protects those with resources from those without (this will be returned to in Chapter 5).

Theories of postmodern culture as a replacement of the social by the signifying nature of consumer culture argue that consumption is organised by lifestyle as opposed to traditional 'ascribed' identities or modern structural divisions. However, differentials of wealth and power can divide society and hence levels of consumption. Moreover, this ignores how gender, ethnicity and sexuality affect our experiences of the world, and how such categories are constructed around differential positions of power, influence and the everyday experience of prejudice, discrimination and fear of violence that pervades everyday life. As such, modes of consumption entirely different from the simple depictions of middle-class-dominated postmodern consumption structure the everyday lives of many people (see Chapter 4).

This section has shown that debates on postmodernism provide varying depictions of the role of consumption in effecting a new social system. What follows in the remainder of this chapter is an assessment of how these often abstract theoretical visions of postmodernism are played out in the spaces and places of the postmodern city.

## Archetypal structural and socio-spatial transformations of the postmodern city (1975– )

The much-debated onset of postmodern conditions, posited around 1975–1979 and characterised by a global oil crisis, economic restructuring and by symbolic events such as the demolition of the Pruitt-Igoe housing estate, led to fundamental changes in many cities. Change manifested itself in a cultural and geographical shift from suburban shopping malls to urban mixed-use developments in the 1980s, when offices, city-centre shopping, entertainment and housing dominated the symbolic economy of the city. While renewal of interest in central urban locations can in part be considered as a new investment cycle, it also reflected institutional and structural changes as financial institutions and information-rich activities expanded on the existing urban base (Zukin 1998a). This was underpinned by the proximity to the city's cultural amenities, which satisfied the needs of professional, high-income wage-earners, so-called 'yuppies', whose salaries and bonuses allowed them to pay high prices for consumer goods, consumption spaces and gentrified housing. There appeared a new vitality to urban life. However, the yuppie lifestyle was widely considered by both 'old' middle- and working-class cultures to have no endearing traits, and those spending freely on high-status goods and services were often characterised as single-minded and self-centred. The speculative investments associated with this period were often blamed for increasing land values, and although this archetypal group had no particularly obvious boundaries of gender, race or ethnicity, yuppies were often blamed for displacing older, poorer or 'industrial' urban residents (Zukin 1998a: 831).

However, yuppies were merely a new wave of gentrifiers of the urban consumption hierarchy movements who, in moving back to the city, endorsed the city's social diversity. This also represented a cultural movement away from the private life of the suburbs and a negation of the historic separation of home and work dating back to the nineteenth century. A desire for authenticity, and large residential spaces at prices the middle class could afford, now became represented in newspaper and lifestyle magazines, where discourses of the aesthetic values of historic homes and lofts glamorised the lifestyles of people in town houses and converted factories. The public image of gentrification was of an 'artistic' or 'bohemian' lifestyle (Zukin 1982), with yuppies buying into and colonising the industrial areas which artists, teachers, writers and creative workers in advertising or retail initially reclaimed from dereliction. Cultural consumption grew with gentrification, with unemployed and underemployed artists and performers working in the food shops, restaurants and galleries. Importantly, the presence of media, music and art led to the development of a critical infrastructure, which promoted the city's emerging symbolic economy in the most successful cities in the urban hierarchy (O'Connor and Wynne 1996).

It is also important to understand how such vitality made urban neighbourhoods more 'interesting' to the middle classes. The nature of middle-class sensibilities demanded high-class private education and thus a movement away from public institutions (see Case study 3.4). Moreover, although gentrification often endorsed social and cultural diversity, this became transmuted into an aesthetic demand for visual coherence of buildings and urban spaces. The advancement of such consumption spaces not only economically displaced lower-income residents who couldn't afford higher rents and taxes, but also culturally displaced the long-term residents through the proliferation of exotic restaurants and wine bars. In essence, the advancement of such consumption practices related to an urban middle-class lifestyle, and a particular aesthetic set of images and forms of sociability, considered as tribalism by Maffesoli (1992).

## Case study 3.4 **The state, regulation and consumption**

An important area of analysis that can be developed in order to address the relationship between social inclusion and exclusion and consumption is the role played by local and national states (Marsden and Wrigley 1995; Wrigley and Lowe 1996). For example, Clarke and Bradford's (1998) discussion of the regulation of consumption questions the relationship between state and private capital in order to assess the ways in which consumption influences the operations of the state, either directly or indirectly. Examples of direct intervention include, for example, the part played by Ronald Reagan's US Republican and Margaret Thatcher's UK Conservative governments in the 1980s – which oversaw the fight between those who saw the future of the economy as, respectively, being service-led or manufacturing-led. Policies aimed at reducing the power of the trade unions, ending industrial subsidy, and privatisation of service provision. An ideological and fiscal goal of this struggle was to reduce the levels of tax levied by central and local government in order to gain the support of middle-class voters to the detriment of marginalised social groups and a social justice agenda.

This is represented in the support for huge spending on spectacular flagship developments such as the Docklands in London over spending on health and education. Moreover, more indirect interventions include, for example, the promotion of new individual rights to consumers, such as food labelling and hygiene (Guy 1996). Such political posturing prioritises support for particular social groups, lifestyles, identities and consumption practices over others.

As noted in Chapter 2, changes in the political economy in the 1980s and 1990s (which chart a move away from the welfare state (*collective consumption*) towards more entrepreneurial modes of governance and state structures) have often been presented as more or less ubiquitous in the urban studies literature. However, the ideological and political outlooks which determine the extent to which collective versus individual consumption is pursued, the extent to which groups of consumers are empowered/disempowered, and how these processes are embraced or resisted at local levels, are of great interest. In sum, tying together the contradictions and tensions between public–private capital, regulation and urban change needs more attention.

For example, Harvey (1989a) suggests that modes of governance that move from a focus on the provision of services to more entrepreneurial government seek to support *private* modes of capital accumulation. Such outlooks, in attempting to make cities attractive to global capital by pursuing urban regeneration projects,

continued

mask the real problems of industrial decline, unemployment, poverty, and social and spatial polarisation (Harvey 1989a; Logan and Molotch 1998). In contrast, postmodern accounts suggest that there *is* solidarity involved in spontaneous structurations of individual agency, and there is considerable potential for collective action through the 'mobilisation' of consumption cultures (Featherstone 1991; Maffesoli 1992; Shields 1992a).

In reviewing the above two positions, Clarke and Purvis (1994) highlight how a more progressive and nuanced understanding of urban consumption can be theorised. They suggest that, whether working at the level of the individual, the city or anywhere beyond or in between, consumption and the regulation and promotion of particular consumption cultures relate to very specific, discursively constructed practices and processes. Thus, while questions of citizenship and consumerism have become important to the (seemingly growing) contradictions between individualisation and privatisation – which are assumed to surround new urban politics and lifestyles – the local 'embeddedness' of constraints and opportunities, and the construction of value systems, have a considerable bearing on how these forces are played out in different cities and different urban spaces and places.

The tribalism represented by Maffesoli (1992) is a cultural movement away from the alienated, private lifestyles of the suburbs and a desire for authentic urban experience. Often represented in the consumer world of lifestyle magazines, movies and television programmes, the urban core again became the centre for a middle-class invasion (particularly young single people, or child-free couples). An ensemble of urban consumption activities from housing to shopping and the proliferation of cultural amenities ranging from restaurants to art galleries developed. However, this was very much a private-sector-led model of urban regeneration, which was based on privatisation of things like housing, education and what has been described as a retreat from the welfare state to more entre-preneurial local authorities pursuing urban regeneration initiatives and domination of urban spaces and places (see Chapter 7).

In order to fully elaborate this new sociality, Baudrillard's (1988) depicted appeal of Disney's consumption regime and its effect on Los Angeles must be considered. The safe, clean public spaces where strangers trust each other in the pursuit of 'fun' inspired some cities to 'Disneyfy'. This included the sponsorship of urban festivals and the development of themed shopping districts. These were policed by poorly paid private surveillance workers. Importantly, these urban conditions maximised the egalitarian values of individualism, autonomy and civic pride, and economic and cultural accessibility to particular spaces/places determined by age and social class became institutionalised in public life.

Urban redevelopment thus became based on these types of social relations and consumption practices, including an array of consumption spaces from restaurants and tourist zones to museums and other cultural activities, casinos, sports stadiums, and specialist stores (see Case study 3.5). In sum, closely aligned with the emergence of this 'new' aestheticised commodity culture was the appearance of 'new' urban spaces. The idea of the aestheticisation of everyday life suggests that a constituent of consumer culture is to produce an altered world, an enhancement of the material environment in such a way as to simultaneously numb and excite the senses. This process of aestheticisation has also culminated in the development of 'total environments', such as theme parks, tourist enclaves, branded shops, bars and restaurants – small, enclosed, self-contained worlds which overload the senses, and where those people with the resources can fashion their own personal and political identity.

These consumption spaces sought to revitalise urban consumption by dramatising the retail 'experience' by capturing shoppers' imagination by inviting them to participate in simulated forms of non-shopping entertainment. Examples include Niketown (see Case study 3.5), the Disney Store, the Gadget Shop, and shops selling outdoors and camping equipment and clothing such as 'wilderness' (REI Trekking) and 'nature' (The Nature Experience). Although these spaces are described as 'entertainment retail', their focus is often to sell recognisable brand names such as Disney, Nike, Sony, Timberlake, and so on. These are also many other examples of new retail spaces such as themed restaurants and bars, including TFI Fridays, McDonald's, Hard Rock Café, Planet Hollywood and Starbucks. There is also a plethora of other themed consumer spaces, and Egyptian, medieval, neo-classical, Roman, rococo, baroque, mock Tudor, the Taj Mahal, Arabian Nights or Wild West Fantasies are branding concepts utilised in a whole range of places from shopping malls to restaurants, bars, shops, boutiques, nightclubs and gyms (Gottdiener 2000). Such new forms of consumption spaces have spread to cities throughout the world from Europe and North America to Japan and China and the city-states of South-East Asia.

## Case Study 3.5 **Commodities and new urban spaces**

The development of the postmodern city is argued to be underpinned by the simultaneous growing importance of commodification of goods and services and

continued

emerging new spaces where they could be purchased and experienced. For example, the first Niketown store which appeared during the 1990s in Chicago was more than an average shop – it was a temple dedicated to the Nike trademark. With its central location on Michigan Avenue's Magnificent Mile and its grand grey marble entrance filled with sporting photographs, the store was associating itself with a glamorous world inhabited by the array of Nike-wearing professional athletes (from a number of different sports) who were recognised as the world's best. In the specific context of the Chicago shop much of the iconography was dominated by images of Michael Jordan, and as the centrepiece of the store, the basketball section was decorated with a 30ft-high photo of the local hero, a mini basketball court, and hidden speakers playing crowd noises.

Other attractions included a genealogy of the sports shoe from 800 BC to 1971 when Bill Bowerman founded the company. The Nike slogan 'Just Do It' was an omnipresent feature of branded areas such as the tennis section, decorated with images of the All England Club and Wimbledon colours. On the top floor of the store was the T-shirt gallery featuring a diverse product range that included Nike Hip-hop T-shirts, Nike Rasta T-shirts, Nike Fifties T-shirts and T-shirts with the classic Nike logo. In an adjoining room a widescreen television continually played Nike commercials. The Niketown shopping experience was more than a shop, it was an event – and in its early days Niketown was the number one sight to see in Chicago, offering juxtapositions that allowed individuals to become playful consumers and to revel in the abundance and choice that were on offer.

One of the most notable of the new consumer spaces over the past fifteen years emerged from the popularity of coffee as it became an increasingly commodi-fied product. Laurier and Philo (2004) show that during the 1990s the rise of cappuccino culture and the ubiquitous presence of outlets such as Starbucks, Costa Coffee, Café Nero and Aroma was remarkable in its speed and extent. In the UK, for example, the number of coffee shops grew from around 300 in 1997 with a turnover of £51 million to 1,850 outlets in 2001 – which was a six-fold increase in five years. Moreover, Laurier and Philo show that such cafés expanded beyond traditional high-street locations and often merged into other venues such as bookshops, do-it-yourself stores and even building societies (as well as airports, and train and bus stations).

Laurier and Philo (2004) considered the importance of this burgeoning 'cappuccino community' and its role in civic life and daily custom in the contem-porary city. They also considered the type of people who frequented coffee shops, and how, why, and with what sets of codes of conduct people inhabit these consumer spaces. They describe commercial research carried out by Costa Coffee that pointed towards four important and distinct aspects of social life that could be found in their cafés. First, coffee shops are popular as part of a quick-break culture where fewer people have time to take long lunches. Secondly, coffee shops provide support for 'multi-tasking' (working, telephone conversation, using laptop computers) in a way that restaurants and fast food outlets could not accommodate. Thirdly, coffee shops are welcoming places for single people and women (more than other venues such as pubs and restaurants). Finally, coffee shops offered a space outside work and home, a 'third' space.

It is clear, then, that bound up in the increased commodification of coffee was a coffee shop 'experience' that is diverse and differentially constructed. The coffee shop is a place to relax, to be alone, to socialise, to read, to gossip, to meet people, to debate, to plan, organise, write, draw, think, vegetate, prevaricate, hide, chew over, swallow, digest and ruminate. It is through such usage that Laurier and Philo describe cafés as 'generative nodes' where economic, political and cultural matters run up against themselves in multiple ways amongst different groups that use them.

Source: Lury (1996) and Laurier and Philo (2004)

Thus, in the 1980s, consumption spaces developed dual aims of stimulating both domestic and global tourist markets. These led to more elaborate shopping centres, retail experiences and shopping facilities – with entertainment to capture shoppers (Zukin 1998a: 834). What follows is a review of some of these Disneyfied and privatised postmodern urban spaces, but also those spaces associated with the 'have-nots' of consumer culture.

## The spectacular shopping mall

The spectacular shopping mall is the epitome of festival retailing, which developed during the 1980s and 1990s. These diversified shopping malls elaborated on the successful regional shopping centres built in the US in the 1950s, and quickly became a dominant force in retailing, competing with the traditional high street and drawing businesses, offices, leisure facilities and retail outlets into privately regulated and controlled environments. Importantly, however, unlike in their suburban predecessors, the inner faces of shopping malls were dramatically re-invigorated, merchandised and repackaged, and appeared in their largest and most spectacular forms in North America, Europe, Australia and Latin America.

This new wave of commercial malls represented a coming together of cultural, demographic, social and spatial factors associated with the postmodern city. For example, architectural designers imploded traditional concepts of urban form by managing to gather together all of the social amenities and shopping experiences of the 'traditional' downtown city street to the suburbs, by playing with space, light, representations and perceptions of safety. City streets, squares, plazas and markets, in many respects the defining sites of urban circulation and exchange, could be carefully miniaturised within the walls of the shopping centre and re-imagined as idealised public spaces, free from the inconvenience of the weather, traffic pollution and unwelcome juxtapositions constituted through the presence of poor people or 'threatening' ethnicities.

While new spectacular shopping malls extended the cultural significance of suburban shopping, integral to this was the architectural advantage that malls promise – a safe, privatised, highly controlled version of the crowded street, free from 'contamination' and benign disorder. One of the best exemplars of this new breed of shopping malls is the West Edmonton Mall, Canada. Through its design and architecture the mall offers a world-within-a-world and is a fantasy landscape juxtaposing a historic present and past through themed streets. While the Mall remains a fortress and protection from the urban disenfranchised – a suburban centre – its historical simulations and artificial representations provide Christopher Columbus's *Santa Maria*, a nineteenth-century European boulevard and New Orleans's Bourbon Street. West Edmonton Mall is a grab bag of submarines and palm trees, baby tigers and skating rinks with layers of history, past, present, future or imagined, with stories of New Orleans black jazz musicians and prostitutes, and dizzying iconographic displays in the middle of a southern prairie city. Bourbon Street is a fantasyland, the world's largest indoor amusement park, with dancing, water fountains and Our Bourbon Street, a corridor of bars and restaurants. Here the Mall becomes a more specific leisure environment, more 'city' than simply interior street. With wrought-iron balconies, arched doorways and shuttered windows, Our Bourbon Street has a repertoire of urban character-istics (Williams 1982).

The key to the success of these malls is that they epitomise suburban values and could be replicated in any part of the city and constructed to meet the needs of middle-class consumers. They provide a safe, privatised, highly controlled version of the crowded street, free from contamination. In this sense, then, the city has relocated inside the stockaded walls of the shopping mall, and the construction of downtown shopping malls in many cities has served to intensify the segregation of the shopping experience. Those whose market position disqualifies them from participating in the postmodern consumer citadels are economically and spatially excluded from them. Shopping malls are strongly

**Figure 3.4 'Europa Boulevard' in the West Edmonton Mall, 1992. (Courtesy of Anne Friedberg)**

bounded and purified social spaces that exclude a significant minority of the population and so protect patrons from the moral confusion that a confrontation with social difference might provoke, reassuring preferred customers that the unseemly and seamy side of the real public world will be excluded.

## Urban villages and loft living

In the postmodern period, the city is framed as vibrant, cosmopolitan, entertaining and 'happening' (Mellor 1997). Hall and Hubbard (1998) describe the frenetic activity of post-industrial urban transformation as ubiquitous through the western world. In these terms 'Image is Everything': the cultural assets of the city are being harnessed, the cultural and creative industry vigorously promoted, and cities are seen as innovative, attractive places to live. Promotion of both high and low culture is undertaken as city 'imagineers' appeal to the consumption practices of the emerging *nouveaux riches* of the professional, managerial and service classes (see, for example, Gold and Ward 1994; Hall and Hubbard 1998; O'Connor and Wynne 1996). Art, food, music, fashion and dance are promoted in the urban 'shop window' (Hall and Hubbard 1998: 199). Images of conspicuous consumption identify the city as a 'fun place to live and work'. In recognition of the complex plurality of the contemporary city, more prosaic 'low' and street culture, working-class traditions and ethnic celebrations are increasingly commodified into narratives of place promotion (Urry 1990; Hall, T. 1995; Zukin 1998a). Cities pay lip-service to the notion of multicultural tolerance of difference (within reason). In the post-industrial economy, then, the symbolic framing of culture becomes a powerful tool as capital and culture intertwine (Jessop 1997; Knox 1987; Harvey 1989a), and it is in urban villages that these processes are most easily identifiable.

Created either through the enhancement of historically distinctive areas, or by developing and generating signatures for previously economically, culturally or spatially ambiguous areas, urban villages or quarters seek to appeal to the consumption practices of the emerging *nouveaux riches* of the professional, managerial and service classes. This is graphically shown by Sharon Zukin (1982) in her famous study of the development of loft living in SoHo, New York. This exemplar of the growth of an urban village unpacked the political and cultural economy of the consumption-led gentrification of a run-down and abandoned area of the city. Zukin describes the colonisation of old industrial buildings that enabled and sustained an infrastructure of musicians, artists, craftworkers, entrepreneurs and cultural producers. These 'urban pioneers' created a vibrant scene which was youthful, dynamic and attractive to particular market segments of consumers and producers who in turn attracted other creative and entrepreneurial people as well as middle-class professional and managerial gentrifiers (see also O'Connor 1998; O'Connor and Wynne 1996). This final group, however, brought higher house prices, concerns with visual standardisation, clean streetscapes and expensive shops, thus displacing the initial wave of gentrifiers to other parts of the city.

Urban villages and quarters are now ubiquitous in cities throughout the world. Central to the vitality of new city spaces are the social identities, lifestyles and consumption practices of managerial, professional and service classes. Similarly, the presence and promotion of, among others, lesbian and gay, youth, and ethnic social groups help to create a vibrant atmosphere in the city centre. This is an urban renaissance based on wealth creation associated with consumption (and the production of consumption), and the cultural and service industries, with a focus on visual attractions which encourage people to spend money, including an array of consumption spaces – restaurants, museums, casinos, sports stadiums and specialist and designer stores (and not traditional industry and manufacturing). This is a post-industrial economy based on the interrelated production of such economic and cultural symbols and the spaces in which they are created and consumed. As such, 'sociability, urban lifestyles and social identities are not

**Figure 3.5 *Loft living, 2004. (Courtesy of* The Independent *and Ken Mackay at Hurford Salvi Carr)***

only the result, but also the raw materials of the growth of the symbolic economy' (Zukin 1998a: 830).

In sum, the economic and cultural vitality of cities is founded on the provision of consumption spaces that include the broadest variety of restaurants, theatres, shops, nightclubs, and so on (Crewe and Lowe 1995). The presence of such consumption spaces is vital precisely because they represent cultural and economic success. In general, the most successful cities contain the most culturally and socially diverse and innovative spaces of consumption. Hence, more successful cities attract a broader range of capital and investment, tourists, and visitors, and in turn attract other innovative and entrepreneurial people: the symbolic success of cities is central to creating a sustainable broad economic base.

## Social control and the postmodern city

The above examples of spectacular shopping malls and urban villages highlight that in the postmodern city the borders between different modes of consumption are heavily guarded and policed, both privately and by the state, and through consumption practices and infrastructure. This is indicative of an increasingly socially and spatially polarised city, divided by areas designed around the needs of middle-class consumers on the one hand, and the poor on the other. They represent a clear example of an increase in the regulation of space in the city and imply new forms of urban sociality and experience. Yet such spaces and places are clearly designed for specific social sectors of the population (or markets). Cities are thus transformed by the need to generate urban conditions that are attractive to tourists and other international migrations of people (for example, businesses, conference-goers, sports fans and those attending cultural festivals), middle-class consumers, and gentrifiers who work in professional and financial services.

Perhaps the most high-profile theorists to consider urban change associated with postmodernity are a group of geographers known as the Los Angeles School. Dear (2000), for example, describes 'edge cities' as one of the archetypal forms of postmodern urbanity. As noted earlier in this chapter, edge cities are master-planned 'suburban' developments with town centres, public squares, police and fire stations and other consumption facilities and activities previously only found in cities, and commercial functions that become decentralised due to high land costs and the social and physical decay associated with inner-city areas. Edge cities grew up around the dominance of the automobile and the communications revolution, and Dear describes three types of edge cities – Uptowns (peripheral pre-automobile settlements that have been absorbed by urban sprawl; Boomers (classic edge cities located at freeway intersections); and Greenfields (occurring at the intersection of several thousand acres of farmland and one developer's ego). Dear also identifies that a key feature of an edge city is that political institutions have not been established but a 'shadow government' exists that can tax, legislate for, and police the community, although it is rarely elected, accountable or subject to constitutional constraints. Dear describes edge cities as reflecting a wider trend towards privatisation, where citizenship is based upon private property ownership, to be protected through security and surveillance, and policed through strict rules about how houses and gardens are to look and be maintained.

Mike Davis (1992) describes such urban spaces as contributing to the creation of a 'fortress city', and further that an obsession with surveillance and separation transformed Los Angeles and the broader region of southern California into a fortress. Davis argues that spatial segregation took place, with residential districts being divided into fortified cells of affluence on the one hand, and places of terror where police battle on the other. Davis (1992: 155) shows that the dynamics of fortification includes high-tech policing methods (helicopters, CCTV and satellite surveillance), security firms operating in gated residential developments and surveillance and policed shopping malls. Davis argues that in the postmodern city the working poor and destitute are often removed from the main streets and excluded from the affluent 'forbidden' cities.

In seeking to conceptualise urban change in Los Angeles between 1965 and 1992, Ed Soja (1996) attempts to link such emergent patterns of urban form to under-lying social processes. He identifies six kinds of restructuring, which together define the region and the city as a 'theme park'. First, Soja describes how the Los Angeles suburbs are promoted as places where the American dream is most easily realised. In an extreme example Soja shows how Orange County, southern California, is a simulation, a structural fake, an enormous advertise-ment, yet functionally the finest multipurpose facility of its kind in the country. He described such urban spaces as 'exopolis' or a city without – a simulacrum of an original that never existed in which image and reality are spectacularly confused. Secondly, Soja notes the development of 'flexcities', associated with the transition to postFordism, especially de-industrialisation and the rise of the information economy; thirdly, 'cosmopolis', referring to the globalisation of Los Angeles in terms of its emergent world city status and to its internal multicultural diversification. Fourthly, Soja describes the city as a 'splintered labyrinth', in which there are extremes of social, economic and political polarisation. Fifthly, there is the 'carceral city', which refers to the incendiary urban geography, brought about by the amalgamation of violence with police surveillance; and, finally, 'simcity' – a term Soja uses to describe new ways of seeing the city that are emerging specifically in Los Angeles, based upon the idea that the city manifests all of the above conditions and spaces, and that as a whole the city itself foregrounds postmodern political, economic, social, cultural and spatial forms.

In contrast to planned or institutionally developed edge cities and urban villages, then, there are a host of other areas, districts or neighbourhoods that must be added to the kind of spaces and places in the postmodern city. Ghettos, red-light zones, neighbourhoods and areas where there are concentrations of marginalised groups and activities, often divided along the lines of class and ethnicity, are bound up in the political, economic, social, spatial and cultural forces which have led to the proliferation of the kind of urban villages outlined above.

## Ghettos, gated communities, defensible space and surveillance

During the early 1980s, particularly in US cities, a complex logic of segregation emerged through both realities and representations of dramatic increases in urban crime and violence. This urban conflict is associated with an intensification of material inequalities and/or racial, ethnic, class or cultural divisions. Fuelling this situation was a fear about unwanted or unpredictable juxtapositions which led to a crisis in confidence over public spaces and re-ordered the heart and environments of many cities.

It was in this context, and in an ever-increasing global movement, that the middle and upper classes in cities are opting to live, work and shop in privately guarded, security-conscious fortified enclaves. This attempt to de-intensify urban space was producing the kind of postmodern 'fortress' cities riddled with sharply demarcated privatised walled and gated enclaves introduced earlier in this section (Davis 1990). They attracted those who wanted to distance themselves completely from the city, desired a socially homogeneous, ordered and manicured environment, and were willing to pay for private services and amenities.

The logic is that once a neighbourhood's boundaries have been cleaned up and streets taken out of use by the public, entry controls can be maintained through a variety of surveillance strategies ranging from increased public policing through to private security guards, to volunteer 'street watch', vigilante or paramilitary patrols. Such neighbourhoods are often enthusiastic supporters of dusk-to-dawn curfews, of identity checks on young people to deter delinquency, gang formation, loitering, truancy, under-age drinking and joyriding, and also of the power to evict neighbours who are considered 'antisocial'. Central to the creation of such spaces was securing them from unwelcome interlopers. In the UK, in particular, there has been a growth in CCTV as a technical fix for problems of crime and security. CCTV has been installed in the central areas of all major cities in the UK. The advocates of CCTV argue that cameras significantly reduce levels of crime, generate a feel-good factor in law-abiding citizens, and allow the police to track criminals and undesirables as they walk the city streets (Fyfe 1998).

The antithesis of the gated community is the inner-city ghetto. Often focused on the decaying modernist housing estates and projects, the inner city has been described as a 'sea of despair'. The ghetto is portrayed as a zone abandoned by formal economic mechanisms and social control and regulation – often becoming a no-go area for the police. These, it is argued, are replaced by forms of informal economy and social control based on violence and threat – with a high proportion of inner-city males likely to end up in prison. Perhaps the most prominent images to be associated with ghetto life focus on social and racial disturbances such as those in South Central Los Angeles following the high-profile police beating of Rodney King in 1992. Such racial disturbances, however, have recently been seen in the UK, France, and Australia (Pile *et al.* 1999).

What is clear, then, is that the inner city has come to represent the 'dark side' of the postmodern city and its decay and criminality a clear juxtaposition with the spectacular redevelopments of the city centre. Mike Davis (1990) is perhaps the most vivid commentator to describe the horrors of life in the postmodern city and the dark side of the postmodern metropolis. He reports on the despair of the post-industrial underclass largely defined by class, gender and ethnicity. He describes a city characterised by wealth and homelessness, divided against itself, and organised along fault lines that separate suburban enclaves from inner-city slums. He shows how the freeway allows middle-class urbanites to navigate the city as a whole without having to encounter the lives of the inner-city neighbourhoods. Davis argues that postmodern 'pseudo-public' spaces such as spectacular malls and urban villages write off the urban underclass, and that the built environment contributes to social and spatial segregation where pariah groups, whether they are poor Latinos, young black men or elderly homeless, are excluded from urban life. Thus, for many the publicly subsidised 'urban renaissance' has produced only a corporate city characterised by armed response, the destruction of public space, and the proliferation of an exclusionary middle-class version of urbanity. The spectacular sites and sights of the postmodern city are nothing more than a 'carnival mask' (Harvey 1989a), eclipsing the increasing poverty and social injustice that have become further institutionalised in individualised postmodern consumer culture.

However, Zukin (2000) notes that despite the many problems associated with increased social and spatial segregation of the urban poor, the diverse and hybrid consumption cultures that exist in poor ghetto areas are resilient and sustainable. Zukin (1998a) argues that shopping streets frequented by immigrants and native-born minorities are central to new urban and ethnic identities. On the streets of New York City, Los Angeles, Atlanta or Toronto the shoppers, peddlers, store owners, managers and clerks are likely to be Africans, Caribbeans, Koreans and African-Americans. These shopping streets create a new African-American identity by interaction among, and fusions between, various traditions of the African diaspora. Although Asians tend to live separately from other minorities, and increasingly in the suburbs, they are active in these shopping streets. Storefront telephone and delivery signs feature many languages, with prices of service to many lands. Newspaper stands owned by members of one immigrant group sell newspapers in different languages. With store owners stocking distinctive ethnic goods, Zukin argues that, on the street, diversity thrives.

Importantly, Zukin also shows that during the real-estate recession of the early 1990s major shopping streets in New York's immigrant districts did not suffer from more vacant shop fronts than those in high-rent districts of Manhattan. On Flatbush Avenue between Church Avenue and Hawthorne Avenue in Brooklyn, 11 per cent of 191 stores were closed, empty or for rent. Store owners were Haitian, Chinese and Greek. Customers were mainly Caribbean blacks, with some Latinos, African-Americans and Chinese. On a Saturday there were no white shoppers or walkers. On Main Street in Flushing, Queens, from Highway to Stanford Streets, 12 per cent of 149 stores were closed, empty or for rent. Most owners were Chinese, but many of the businesses were corporate franchises. On a recent Saturday mainly Asians were on the street, with some Latinos, Indians, African-Americans and whites. However, despite this economic segregation, vacancy rates were similar on Madison Avenue in Manhattan, where storeowners and customers were predominantly white and often European. In 170 stores between 80th and 96th Streets, the vacancy rate was almost 10 per cent. During the 1970s, vacancy rates were 15 per cent. Similarly, between 57th and 69th Streets, where there was a high concentration of European designer clothing and luxury goods stores, the vacancy rates, of just under 10 per cent, were the same as in Flatbush.

## Disrupting archetypes of the postmodern city

Earlier in this chapter I questioned the extent to which mix-and-match postmodern consumer culture had replaced social differentiation through categories such as class, gender, ethnicity and sexuality (this is returned to in Chapter 5). This argument is further elaborated by Wynne and O'Connor (1998), who argue that the very making of culture itself, the legitimation practices it produces and its aesthetic content in specific socio-spatial settings must be addressed. As such, consumption (like other cultural forms) must be considered as *the means by which social structure is mediated to and by individuals*. Consumption cultures, like other cultures, provide the stuff that allows such mediation. However, if culture and consumption are so important to urban change, and cities and social groups are so diverse, can these processes really be generalised? Does theorisation about the postmodern city really help us to understand urban change as experienced by all urban dwellers and in all the city's spaces and places?

It is perhaps understandable that consumption has predominantly been considered in general terms (although often through particular case studies), as part of a project to ground it in particular master narratives of particular periods. Theorists have linked diverse political, economic, social and cultural processes, which have provided a valuable shorthand to describe urban change. Such advances now allow the heroes and villains and the bright lights of glamorous cityscapes to be a staple focus for academic study as well as pervading popular culture. This has been particularly important in crossing the boundaries between conceptualising the economic and cultural shaping of material life and the moulding of human desires.

Nevertheless, there has undoubtedly been a persistent problem of abstraction. Consumption has been glossed over as a composite and synthetic term. It is at once central to arguments about long-term economic transformation, and to the shifting contours of political discourses and re-ordering of space, place and identity without fully presenting the variable ways that these practices and processes are manifest in particular places at different times (Wynne and O'Connor 1998). For example, the suggestion that certain spaces of consumption and certain commodities can be associated with certain classes is often overstated. In everyday life it is not such a leap of the imagination to see not only that people from different social groups consume a significant proportion of the same or similar goods and services, but also that they co-exist in the same shops, supermarkets, malls, streets and other urban spaces and places (Crewe 2000).

Furthermore, while consumers may be pursuing what appears to be an aesthetic-isation and stylisation of everyday life (implying a detachment which seems to discourage sympathy with other urban groups as well as pleasure; see Featherstone 1991), this does not necessarily translate into political disengage-ment or polarisation. As such, generalised depictions of a retreat from the welfare state, aestheticisation and standardisation of the urban landscape, associated with a collective abandonment of collective experience, have overblown not only the connection between private consumption and aestheticisation of an 'anti-urban landscape', but also the very extent and constructions of these processes themselves (Davis 1990).

Any measure of the advancement of cities (and spaces/places within cities) must appreciate more fully the discursive construction of changes in types of sociability, social identities and lifestyles. It is also clear that these occur in different sequences, in different localities, to different extents and with different characteristics (Appadurai 1996: 71–73). There is a need to look at the way urban change, economic and cultural strategies and urban regeneration are rooted in local imaginaries in particular places. Such an approach seeks to identify micro-geographies, and to present the 'hybridity' of (post)modern social identities, and to promote geographies which conceive social relations as dynamic practices forged anew by creative bricoleurs using space creatively in different places (Jackson and Thrift 1995).

For example, while many of the new consumption spaces rely on a high level of skill and knowledge, and provide cultural products of beauty, originality and complexity, others are standardised, trivial and oriented towards predictability and profit (Ritzer 1998). At the same time, individual men and women express their complex social identities by combining markers of gender, ethnicity and social class. As such, study of cultural styles, lifestyles and identities created and diffused in cities streets – made by people who work in urban environments (artists, single parents, designers, feminists, people who belong to gay and immigrant cultures) – provides valuable understanding of their specific location.

Hence, the ubiquitous shopping street or cultural districts which have been shown to be frequented by immigrants and native-born minorities, or by those pursuing the postmodern mix-and-match lifestyle, must be understood as an integration and fusion of different local traditions and diasporas, and their construction in terms of specific socio-spatial characteristics must be addressed (see, for example, Bhabha 1994). The presence (or not) of spaces and identities can thus tell us a lot about the locality within the construction of local, regional, national and supranational economic and political conditions, and social values which make city spaces/places and regions unique and characterful (Zukin 1998a).

This argument allows us to go beyond an interpretation of urban consumption posed in the broadest terms of social and urban theory. This allows us to undertake a critical interpretation of postmodernity which can thus make connections between the production of physical spaces and symbols and between the physical environment, sociability and urban lifestyles, choices and reflexivity (Lash and Urry 1994). It is clear that studies of archetypal geographies of feudal, pre-industrial, industrial and (post)modern cities can be presented by identifying key features of physical, economic, political, social cultural and spatial practices and processes. Nevertheless, 'on-the-ground' geographies are produced by characteristics, social institutions and practices which are very different in different places. For example, presenting a break or dichotomy between the modern and the postmodern city by comparing the social life of industrial production, social administration, commercialisation of leisure and aestheticisation of lifestyles and places/spaces is misleading. Both archetypal (post)modern and industrial class structures can be easily presented as eroded, fragmented and hybridised in specific cities and spaces/places. As Glennie (1998) suggests, the analysis of urban forms can only begin to be more relevantly pursued through careful consideration of the disposition of the material and social fabric and personnel of specific cities (and spaces/places within cities).

Chapter 2 showed that few cities fit squarely into archetypal descriptions of what a modern city should look like. And as York or Durham in the UK, or any other number of European medieval cities, retained much of their pre-industrial structure in a modernist period, the same can be said for modern urban characteristics of people and places in an era of postmodern urbanity. In reality, many cities will demonstrate some combination of modern urban characteristics mixed with newer postmodern urban forms. At local level, things are uneven, resisted, misread and dependent on local embedded social relations.

It is also clear that the distinction between modern and postmodern urbanism is empirically difficult to sustain, as is their periodisation. However, what such epochal archetypes offer is a useful template to be critical about political, economic, social, cultural and spatial practices and processes taking place in cities throughout the world. What is in less doubt is that consumerism has taken over as the logic of social control, with worker citizens becoming consumer citizens, and that has led to an ever-widening gap in economic and spatial segregation between 'haves' and 'have-nots'. However, in many cases the construction and maintenance of such divisions are more complicated, fluid and elastic, and we must be careful how we conceptualise this divide.

For example, there is still an illusion of mass consumption that continues to be cultivated in entertainment-focused retail shops, malls, multiplexes and public spaces, and it is true that in spaces and places within the postmodern city people from different social and ethnic backgrounds are brought together in consumptive activities. However, by the same token new surveillance technologies and policing ensure that the 'seduced' and the 'repressed' – while both being present – can be held apart in such spaces. However, as noted in the introductory chapter, consumption is also a productive force that offers the opportunity for the appropriation and re-appropriation of urban space. As such, while the polarisation thesis is too simple, and it is clear that some consumption spaces are shared, the interaction of the seduced (who can fully engage in consumer society) and the repressed (who take part but not in the ways they would wish) is underpinned by complex practices and processes. This relationship between the 'haves' and 'have-nots' (or the seduced and the repressed) and its spatial imprints are more fully addressed in a number of different ways throughout the remainder of this book.

## Concluding remarks

The genealogy of the people and places that characterise the postmodern city that has been outlined in this chapter allows us to construct a valuable generalisable template. This allows us to identify how broad urban agendas are articulated in the fabric of particular cities, as well as providing a framework for comparative research. However, of equal importance is the fact that consumption practices are now seen to underpin the economic and political life of cities, and so must be recognised as a foundation of forms for social relations, sociability and the nature of urban life itself. Nevertheless, what the stories of both the modern and the postmodern city show is that no one story 'fits all', and our understanding of the relationship between cities and consumption must be posited in both generalist and specific ways, both theoretically and with empirical research.

Learning outcomes

- To understand the relationship between consumption and the development of the postmodern city
- To be familiar with archetypal postmodern identities, lifestyles and forms of sociability
- To be able to describe the new urban spaces and places associated with postmodern consumption cultures and the increasing social and spatial polarisation of the city
- To critically assess theories of postmodern urbanity and to understand nuanced perspectives

# Further reading

David B. Clarke (2003) *The Consumer Society and the Postmodern City*, London: Routledge. A detailed critique which explores with great clarity high-profile and dense writings on the postmodern urban condition.

Mike Davis (1990) *City of Quartz: Excavating the Future in Los Angeles*, London: Verso. This book provides a valuable exploration of an archetypal postmodern city, Los Angeles, and eruditely describes and explains practices and processes involved in the increasing social and spatial segregation of that city.

David Harvey (1989) *The Condition of Postmodernity*, Oxford: Blackwell. Harvey offers a well-rounded and empirically informed analysis of changes in the mode of capital accumulation and the associated social and cultural constituents.

Sharon Zukin (1988) *Loft Living: Culture and Capital in Urban Change*, London: Radius. This ground-breaking book looks in detail at the political, economic, social, cultural and aesthetic constituents of a middle-class return to the urban core.

# 4 Consumption and everyday life

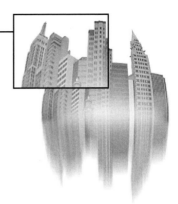

**Learning objectives**

- To look at the relationship between consumption and everyday life
- To show how ordinary and mundane consumption is as complex and rich a topic for study as more spectacular consumption
- To think about the relationship between power and resistance that is played out through everyday consumption
- To describe the relationship between everyday life and consumption in a range of different spaces and places

This chapter acts as an important antidote to the spectacular urbanism outlined in the previous chapter. The focus here is on 'ordinary' or 'mundane' consumption, 'everyday' urban spaces, activities and social relations. This includes discussion of 'inconspicuous' consumption spaces such as car boot sales, charity shops, retro/second-hand clothes shops, markets, supermarkets and the home. Mundane and everyday worlds are shown to have hidden codes and languages every bit as exotic as spectacular spaces, places and formal retailing, if not more so (Crewe 2000). Moreover, the ways in which people engage with the spectacular urban landscapes in ordinary and mundane ways will be highlighted (de Certeau 1984; Lefebvre 1971). This chapter will unpack the ways in which individual agency relates to consumption practices, and thus highlight the weakness in more structurally biased urban studies of archetypal urban spaces/places and identities, lifestyles and forms of sociability.

Studies of inconspicuous consumption spaces have generally been combined with interpretation of practices that constitute identity formation. Studies of authenticity, rituals of re-appropriation, recycling, repair and restoration have suggested that consumption is not a single act of purchase but can be a potentially endless circuit of use and reuse. Similarly, research into domestic and gendered knowledge(s) in everyday spaces of the home, 'virtual' shopping through catalogues, classified advertisements, the internet and cable shopping, speciality magazines and DIY, has identified the symbolic meanings of these activities, and the way in which their (re)negotiation and (re)working represent a productive, creative and reflexive process.

Importantly, studies of consumption at the level of the mundane and everyday have asserted that considerations such as 'love' and 'value' have been ignored. There has been a tendency to oversimplify or ignore exchange relations which are not rational, but are imbued with social meanings and are highly 'embedded' in localities with richly symbolic activities with emotional contexts (Miller 1998). To this end, Crewe (2000) not only asserts the need to map sites of consumption (both spectacular and inconspicuous), but stresses that an anthropological approach exploiting methodologies such as participant observation and field diaries provides a more nuanced understanding of the social significance of contemporary con-sumption practices and their relationship to individual subjectivity. It is through the use of such methods that studies of consumption have pushed back their own boundaries and have addressed topics such as food, fashion, mobile phones and home decoration, as well as the ways in which 'the street' is a reservoir of ideas for catwalk fashion designers (Crewe 2000).

Research focusing on consumption and everyday life has thus been important in several ways. First, in highlighting the potential to move away from a concentration on consumption and spectacular urban landscapes, and the realisation of the important role that consumption plays in everyday and mundane life. Secondly, by stressing that active negotiation, subjectivity and conflict are important in fully conceptualising the discursive construction of identities, lifestyles and forms of sociability. The following section will look at theories of everyday life which have borne such theoretical understanding. This will be followed in the rest of the chapter by a review of how theories of everyday life and consumption have been folded together, followed by brief examples of this synthesis. The second half of the chapter will introduce urban spaces and places that have been highlighted in studies of ordinary consumption and associated constructions of identity, lifestyle and forms of sociability.

However, as noted in Chapter 1, commodities and consumer spaces are imbued with symbolic meaning that can be appropriated and (re)appropriated and hence 'consumed' in many different ways. It is in this context that it can be argued that concepts such as 'mundane' and 'spectacular' are themselves fluid and elastic. In a world where increased commodification dominates everyday life and both individual and social concerns, the interpretation and appropriation of the very conception of what constitutes 'spectacular' and 'mundane' can be seen to be different for different people at different times. For example, city dwellers often engage with spectacular urban consumer spaces in unthinking and ordinary everyday ways. Moreover, even seemingly mundane spaces and consumer activities can take on magical and spectacular characteristic and interpretations. Routine and unthinking consumption and extraordinary interpretation and appropriation of commodities can thus take place in marginal sites and activities as well as in spectacular urban settings.

For example, eating out at an exclusive city restaurant located in a dramatic waterfront development can be as much an everyday experience as eating at home. Shopping at a local market might offer sights, sounds and experiences that are exciting and extraordinary. With this in mind, this chapter, in seeking to discuss the everyday as an antidote to the spectacular urbanism outlined in the previous chapter, does not equate the everyday with the mundane. What follows in subsequent sections is examples and conceptualisations of the diverse and complex ways in which everyday consumption can be both spectacular and mundane.

## Theories of everyday life

The two most high-profile writers engaged in theorising everyday life are Michel de Certeau and Henri Lefebvre, who have both brought attention to the experiential dimension of urban life and the significance of the city as a site of resistance. De Certeau, for example, is concerned to investigate the 'tactical' ways in which everyday activities or 'practices' are utilised by the less powerful to subvert dominant social ideologies and power relations. By describing consumption practices such as reading, cooking and shopping, de Certeau highlights the creativity of everyday life. Importantly, he argues that the resources of a more powerful 'other' are routinely appropriated by and used to service the interests of the subordinate. Such tactics, he suggests, have too often been overlooked or dismissed by urban theorists concerned with more technocratic and systematic conceptions of power and their impact on the development of the postmodern city (de Certeau 1984: xi).

Both writers took as their starting point the idea that at the heart of the concept of 'everyday life' is a definition of 'culture' as ordinary. Studies of everyday life are concerned with the culture of our daily lives – what we all do, all of the time. Hence, studies of everyday life are not concerned with the powerful and that which is recorded and codified. Instead, writers interested in conceptualising everyday life are fascinated by that which is unpredictable. This interest has rested upon the routine activities of ordinary life and what people do as they go about their day-to-day lives. In sum, studies of everyday life are interested in the routines that we repeat daily, the humdrum, the taken-for-granted.

Henri Lefebvre (1991), for example, shows that the production and maintenance of places and spaces are inseparable from the social networks which give meaning to everyday life. He argues that the meaningful social networks that make up our lives are constituted and lived by people as part of 'everyday life'. As such, people produce such networks – through contact with others, establishing frames of reference, the routinisation of everyday practice; people live through these social networks. In short, meaningful social networks and the ways individuals make sense of those social networks (networks of meaning) are both a product of social life and the meanings through which social life is constituted (Gardiner 2000).

However, Michel de Certeau notes that the practices that make up everyday life 'elude discipline'; they remain 'invisible' (1988: 96). In other words, 'these actions take place within existing (imposed) regulatory frameworks but manage to avoid the nets of surveillance, policing and discipline' (Stevenson 2003: 67). Stevenson draws on the work of John Fiske (1989) to elaborate this point. Fiske shows how young people routinely take the objects or commodities of capitalism and use them creatively to express identities and values – such as clothing, magazines and other popular cultural forms. This creative use is frequently in opposition to those of the dominant conceptions of utility. Safety pins worn by many punk rockers in the late 1970s, for instance, were never intended by their manufacturers to be pierced through the nose (Hebdige 1979). However, worn in this way the safety pin became an act of resistance and a marker of collective identity. In a similar vein, Stevenson (2003) uses the example of a rented flat to explain Lefebvre's and de Certeau's ideas – in a rented flat a tenant makes the space their 'own' through the socialising they undertake. This can include their choice of furnishing, the placements of personal effects and the uses they make of the rooms (a lounge room used as a bedroom).

Nonetheless, despite such acts being about resistance, the potency of such tactics to upset balances of power, structures of dominance and subordination identified by Lefebvre and de Certeau must be questioned. For example, Stevenson (2003) shows that the opportunism of capitalism is such that styles created by subcultural groups as markers of resistance are frequently (re)appropriated by mainstream culture. Punk is just one example of a subcultural style that emerged first in an 'underground' movement of musicians, fashion designers, and in fanzines but was later mass-produced and mainstreamed in fashion chain stores, by the music industry and on children's television. Nevertheless, the creativity of popular culture and everyday life means that new styles (such as the 1950s teenager, mods and rockers, punks, ravers, and gangstas) are continually being adopted as tactics of resistance. Similarly, the tenant that attempts to add personal touches to their rented flat (but always within the parameters set by the formal agreement they have with the landlord) in order to claim the space as their own – space that they do not own in an economic sense. However, Stevenson (2003: 67) is right to show that despite such constraints, these tactics can (albeit in a small way) serve to 'disrupt or undermine capitalist social relations and the power derived from the private occupation of property', and thus play an important role in framing and expressing the identity of the tenant.

What these examples show is that both Lefebvre and de Certeau are correct to assert that the city, and the uses that urban dwellers make of its spaces and places, is an important 'site' of bottom-up resistance. Stevenson (2003) suggests that just as Benjamin recognised the contribution of urban walking to the construction and investigation of a culture, walking is central to the work of theorists of everyday life – both as an activity and as a metaphor. Of interest to both theorists are popular practices and the culture(s) of the street and reading the city as 'lived text'. For example, de Certeau shows how the overarching meanings that have been inscribed or imposed on the landscape by official processes or 'strategies' such as architecture, planning and design can be subverted by 'spatial practices' or 'tactics'. In opposition to the rational official 'concept city', that exists as a totalising and knowable space, de Certeau posits an unknowable, dynamic city, the city (as a collection of spaces) unpacked through consideration of its smallest elements – footsteps in the street.

In a similar vein, Lefebvre concurs that the importance of such an approach is to interrogate the view that in the (post)modern era imagination and creative human activity are transformed into routinised and commodified forms. In these terms, Lefebvre suggests that such routinisation is both repressive and emancipatory. His critique of everyday life highlights the habitualised and recurrent activities that make up the mundane character of everyday life. By discussing topics such as desires, labours and pleasures he shows that everyday life is filled with both production and consumption activities. This ensures that, even at the level of the everyday and mundane, consumer society as a 'way of life' (Miles 1998b) is part of our social relationships. In sum, he argues that the habitualised drudgery that makes up our everyday life is firmly centred around consumerism and the social control and ordering of everyday life. Nevertheless, given that there is creativity and individual agency involved in everyday consumption activities, there is – to both Lefebvre's and de Certeau's minds – a certain degree of resistance to such social control and ordering possible in mundane and everyday activities.

For example, through walking, de Certeau argues, urban users both experience and create a city they cannot 'see' – in the sense that it is possible to be only in the space you are occupying at a given time. The route that individuals take (where you have been and are going) through urban space can be held in memory or imagination. This mental map can have been informed by its representation in a map or pictorial form or by previous experience of the particular space. De Certeau argues that it is in the act of walking that a 'myriad' of users write and rewrite the city as 'their' space – creating fragmentary stories that link and intersect with other fragmentary stories (Stevenson 2003). It is also through these trajectories and connections that the city is given form (de Certeau 1988).

De Certeau elaborates on this process and highlights its significance not only in terms of networks of moving and intersecting 'writings', but by suggesting that urban dwellers compose a manifold story that has neither author nor spectatorship. However, while these urban stories are shaped out of fragments of individuals' trajectories and spatial stories in relation to official representations, they remain a daily and indefinably 'other'. Thus, while it is possible to trace or mark a walker's journey on a map, such representations fail to capture the discursive quality or the real nature of the experience of walking around the city. Any such representations are stripped of content and do not fully elaborate the ways of being in the world. In addition, Stevenson (2003) argues that such representations fail to reveal the tactical quality of 'the walk' and the ways space is claimed and interpreted through the act of walking.

For the city-dweller, then, the 'invention' of place through practice also involves the remembering of stories associated with the place that, in turn, contribute to their connections to that place – these are spaces of imagination and place-based experience and nostalgia (Stevenson 2003). For example, de Certeau highlights the importance of physical and symbolic boundaries (such as place names and street furniture) to the process of experiencing and remembering the city, and in constructing the lived culture and experience of the urban street. He suggests that the boundaries which delineate everyday urban experience have both a spatial and a narrative form and serve an important role in delineating one's place in the world – defining who one is as an occupant of cityscapes. In other words, there are spatial cues within the fabric of the urban environment that we either associate with or consider are 'not for us'. In simple terms, these are about inclusion and exclusion, and are employed in the negotiation of identity by people on a daily basis. Frequently, in the absence of personal experience of the place, we impose a dominant impression that is ascribed to it deliberately or otherwise. This is done in a myriad complex ways from affiliation or legibility of particular shops, the absence or presence of particular social groups, types of houses, cars, and street culture to other sights, sounds, smells and consumption opportunities (see Chapter 6).

This process is a significant factor contributing to the development of positive and negative images of place, including the stereotype of the stigmatised neighbourhood, suburb, city or region and in framing imagined urbanism (Stevenson 2003). In addition, the importance of meaning to this process of mental mapping/ stereotyping cannot be underestimated. As such, de Certeau suggests that the official city exists, in part, as an outcome of having been named – the name unifies and sets it apart. This is equally the case with streets and suburbs. However, names (or at least 'proper' names) also give places a meaning that might have been previously unforeseen, and thus a symbolic dimension which 'eludes' the official or imposed. For instance, according to de Certeau (1988: 104) there are a diverse range of possible interpretations and meanings given to names by passers-by. Names can be detached from the places they are supposed to define and serve as imaginary meeting-points on itineraries. There are also those names which linger in popular usage long after 'the place' has gone and which can confuse the casual visitor or newcomer (Stevenson 2003).

In many respects, the symbolic unification of previously unrelated spatial fragments such as different parts of a city, region or nation begins with official and unofficial naming of the space. This asserts that place naming is the act that transforms symbolic anonymous spaces into particular places for which a mythical past can exist and an imagined future is possible (Stevenson 2003). Much is invoked by films, songs, magazines and popular imagination about particular places (such as Manhattan, Manchester, Timbuktu and Baghdad). In this context there have been more images and representations of New York than of any other city in the world (Stevenson 2003).

What writing about everyday life shows is that there is an intricate connection between the naming of places and encoded cultural meanings. In other words, the official representations of the city and of physical streets and buildings unfold hand in hand with both popular representations and ways that individuals write their urban experiences on a daily basis, and the various representations inform each other. However, these spatial meanings are not essential, nor are they fixed or stable. Rather, meaning is problematic, requiring continued definition, redefinition, constitution and reconstitution through discourses including popular culture and activities of walking and occupying space (Stevenson 2003). Equally, spatial meaning, including naming, is vulnerable to manipulation by particular interest groups for a range of ends, be they political, social or commercial (see Chapter 7).

Since the mid-1970s there has been a burgeoning interest in investigating the ways in which people use and experience the city. There has been a particular focus on uncovering the meanings that people derive from this engagement and an interest in culturally informed research. This has focused on the actions and interactions of urban dwellers – life as it is lived on the streets and in the neighbourhoods – and framed analysis through these micro-situations of everyday activities. Work has sought to explore the symbolic and interpretive dimensions of urban life. This has attempted to move beyond the structural, the predictable and the political to delve into the irrational and serendipitous, and hence the themes of the symbolic and the interpretive recur in studies of the city (Gardiner 2000). In this respect, much of this work links with, and makes use of, the insights of several key theorists who have addressed the relationship between the structural and the lived city. What this section has highlighted is that both de Certeau and Lefebvre note the importance of tactics and strategies and how within the context of broader structural urban life such everyday experience makes up the seductive qualities of (post)modern consumerism. This allows us to feel ourselves as individuals who are relatively free to construct our own sovereignty and selfhood, but only in the context of the broader rigorous constructions of social relations control that are manifest at the level of the mundane and ordinary (Gardiner 2000: 168).

## Consumption and everyday life

It has been shown that studies of everyday life have been important not just in rectifying the previous lack of attention to theorising the humdrum, the taken-for-granted, the routines that we repeat daily, but in identifying how consumer culture pervades social relations of daily life. Bound up with this understanding is the work of de Certeau and Lefebvre, who show how 'everyday life' is made up through productive consumption. These theorists have shown us that consumers are almost endlessly creative in the appropriation and manipulation of consumer goods. Importantly, this view is in complete opposition to the 'mindless dupes' thesis promulgated by the Frankfurt School. In simple terms, theories of everyday life have shown that identities are constructed through everyday practices, that goods and services are transformed through creative consumption practices, and that the city is consumed in multiple ways (see Chapter 6).

One of the most important landmarks in the developing literature concerning everyday life has been to posit 'everyday life' as the binary opposite of state bureaucracy. This refocuses attention away from the long arm of state or other impersonal institutions involved in administering laws and regulations or pursuing the economic rationality of market relations (Stevenson 2003). In these terms, everyday life is characterised by the following concepts: small, local communities with close and emotional ties, connectedness between people, caring and spontaneity, immediacy, participation and collaboration. This has proved to be an important antidote to studies of consumption that have concentrated unduly on the more spectacular and visual aspects of contemporary consumer behaviour and urban life, thereby constructing an unbalanced and partial account (see Case study 4.1).

## Case study 4.1 **The farmers market**

In her book *Point of Purchase: How Shopping Changed American Culture*, sociologist Sharon Zukin describes the social, economic and cultural significance of 'alternative' urban shopping experiences. Living and shopping in New York, Zukin describes the pleasures of shopping at an 'on-street' farmers market, in contrast to the spectacular urbanism represented by shopping malls, department and retail entertainment stores. Zukin's appreciation of the farmers market is based around regular face-to-face encounters with people, the smell, touch, and being able to gaze upon the products. As stores have become larger, more pervasive and (arguably) less personal and more spectacular, Zukin suggests that her once-a-week visit to Union Square to seek refuge in farmers markets is because it is a place where she can meet with producers to find out where and how food is produced – she is able to develop personal relations with the people who produce and supply her food.

New York has a citywide network of Greenmarkets that have been a regular feature in the city since the 1970s. The markets reflect the desire of many consumers from different social backgrounds to consume healthy food. Zukin enjoys the opportunity to find high-quality fresh tomatoes from the country in the middle of Manhattan (although not necessarily at prices that are lower than in grocery stores).

One particular shopping treat that Zukin looked forward to was visiting (the now deceased) Mr D'Attlico, the lettuce man, with whom she had developed a personal relationship over many years. Such a relationship was characteristic of the farmers market, and Zukin argues that a key feature of the farmers market shopping experience is an interaction between shoppers who sought advice or just chatted about the produce and topics such as home freezing, levels of preservatives, the length of time fruit takes to ripen, and so on.

Importantly, the farmers market is beneficial not only to the consumer but also to the producers in that only local produce can be sold, and selling direct to consumers allows the farmers to charge retail prices. The regular farmers market is also an opportunity for producers to socialise, share news and exchange produce with one another. For example, the women who sells eggs trades with the woman who runs a large vegetable stall, and the maple syrup man barters with everyone. Zukin argues that after thirty years the Greenmarket remains a magnet for shoppers and, due to its popularity, has been the impetus to regenerate a nearby park, and has

continued

attracted people to live in close proximity to the market. The farmers market is thus a place that generates a specific and diverse community of people. It is an important urban consumption space with particular associated forms of conduct and social relations in the daily lives of many New Yorkers.

Source: Zukin (2004)

This folding together of theories of everyday life and theories of consumer culture is important in asserting the 'pleasures of consumption', and moreover in considering popular culture as a contested arena which involves a confluence of creative everyday practices and products. With this understanding comes the view that consumption is not only empowering but also subversive. This draws diametrically opposite conclusions from the pessimism of the 'mass culture' critique. In sum, consumption is not the end of a process, but the beginning of another, and thus itself a form of production.

For example, Gronow and Warde (2001: 4) suggest that social scientists have concentrated too much on topics such as music tastes, clothing fashions, private purchases of housing and vehicles, and attendance at high cultural performances like theatres and museums. This has led to the exclusion of everyday food consumption, use of water and electricity, organisation of domestic interiors and listening to the radio at home. They argue that there has been too much emphasis on extraordinary rather than ordinary consumption; conspicuous rather than inconspicuous consumption. This has also over-privileged individual choice rather than contextual and collective constraint. Moreover, while conscious, rational decision-making had been considered, routine, conventional and repetitive conduct had been ignored. Similarly, decisions to purchase (rather than practical contexts of appropriation and use), commodities (rather than other types of exchange) and personal identity (rather than collective identification) have all been over-represented in the literature.

For example, there are a whole range of consumption activities that take place in different spaces and places within the city and contribute to the constructions of urban lifestyles, identities and forms of sociability. Ilmonen (2001) talks about routinised and repetitive consumption, such as brand loyalty and the way in which food preferences may be acted out in a totally unthinking way – a product of acculturation and part of the habitus of everyday life. Similarly, Sassatelli (2001) addresses the pursuit of immediate pleasures, and discusses the normalisation of pleasure, drugs and alcohol. Sassatelli shows that drink and drug cultures are played out in places and events such as rave parties, discos, pubs and clubs. It is in such spaces that participants can and indeed must follow certain manners, codes and rules. Sassatelli (2001) describes this as ordered disorder, where dancing at a rave or drinking at your local pub or bar provides time off from everyday reality.

One interesting critique of the role of everyday consumption on urban life comes from Maenpaa's (2001) research into mobile phone consumption. Maenpaa argues that an important part of mobile users' everyday life has been to create a distinct culture. This represents an intensification of interaction, but not necessarily a decrease in the importance of face-to-face interaction. Maenpaa argues that through mobile phone use, public and private intermingle in urban space, and that this represents a public statement of the segregation between intimate and public life. Thus, the use of mobile phones increases the 'liveliness' of public spaces by increasing the amount of noise and conversation taking place on the streets and other public spaces, thereby making urban life more lively, which today is considered an important characteristic of a liveable city. However, although mobile phone use adds to public life, Maenpaa argues that the public sphere is narrowed down as people bring their private lives inside it through a private and (for the casual listener) a one-way conversation or mild annoyance. As such, the mobile and its public use cannot overcome the urban public norm of ignoring other people and avoiding personal contact.

In a similar vein, Dant and Martin (2001) discuss the large-scale physical mobility in the city provided by mass access to transport, and particularly the car. They suggest that, like televisions, telephones, central heating and inside toilets, cars are part of the equipment of modern living that is more noticeable when absent than when present. From joyriding to Sunday afternoon motoring, flows of everyday relationships are formed around human social relations with the car – and sociality extends freedoms at a cost. For example, the contemporary location of shops, schools, places for work, leisure facilities and other services in relation to where people live often makes the car more convenient, if not inevitable. Indeed, in many cases it is the primacy of the motor car that has more determined the physical development of cities over the past hundred years than has human sociality. Without cars, the pattern of consumption of many modern lifestyles would be impractical. Moreover, lovers, couples, families, parents and children all have different forms of sociability in cars – sex, chatting, singing, listening to music, arguments and sightseeing are all undertaken in cars (see Chapter 6 for more discussion on the car).

It is clear, then, that such studies show that forms of consumption are habitual, repetitive social patterns of behaviour rather than of people's conscious, more or less imaginative choices. This represents an unconscious non-reflexively applied routine and reflects how people paradoxically imitate each other in order to feel that they are expressing their individuality (Gronow and Warde 2001: 4).

Consumption practices are shown to be central to even the most mundane and everyday activities, but within the dominance of structures of consumer culture individuals can use consumption practices to construct their identities, and negotiate and experience the world around them in creative ways. What follows in the remainder of the chapter is fragments of ordinary urban consumption. First, there is a focus on the home, in particular on decoration and consumption associated with the kitchen. Secondly, we shall look at eating out and urban foodscapes, followed by a discussion of weekly shopping at the supermarket. Thirdly, consumption cultures associated with second-hand shopping will be discussed. Finally, we shall look at how consumption in old age is constructed through particular social relations and sociability, and how this impacts on the everyday urban lives of older people.

## Consumption at home

The development of the physical organisation of the postmodern city can be argued to a large degree to be oriented around the relationship between where people live and where people work. While there has been a large amount of writing about people's work life and other spaces and places concerned with urban production and consumption, it is only recently that there has been a focus on consumption at home. This interest seeks to develop the idea that the nature of 'home' owes its production as much to social imaginary as it does to material constitution (Crewe 2000). The home remains the largest single economic investment most people are likely to make in their lifetimes, and indeed has an enduring cultural salience as a symbol of modernity, bourgeois respectability, status and success (Stevenson 2003).

It is not simply that home and domestic space are important as repositories of the goods bought from the plethora of high-street stores, shopping malls, super-markets and markets that make up the urban consumption landscape, but that the home is an important element of study in itself. To this end, studies have questioned the creative dimensions of domestic consumption and the ways in which consumer goods are actively appropriated in the everyday spaces of the home. Other areas of interest include shopping at home, through consumption via catalogues, classified ads and Tupperware and Ann Summers parties, and moreover the ways in which discount stores have an aesthetic appeal to social groups excluded from expensive high-street shopping. Such studies have identified how thrift and utility are mediated by a matrix of knowledge, authenticity and the skill involved in finding a bargain (Crewe 2000). Other areas of interest include the rise in new technologies, such as the internet and cable television, and their impact on consumption practices (Lally 2002). The high-profile popularity of lifestyle magazines and television household makeover programmes has also been scrutinised (Holliday 2005).

In the proliferating literature surrounding consumption in the home, there is a growing interest in issues of identity, lifestyle and forms of sociability relating to a whole range of household practices and things. For example, Southerton (2001) discusses kitchens, and argues that because these rooms, often described as the 'heart of the house', are used in everyday life, their symbolic meaning cannot be reduced to material tastes. How kitchens are used offers insights into the social and symbolic role of consumption, and their decoration and use embrace moral categories and judgements of consumption practices.

Southerton uses the example of the kitchen table to elaborate this argument. Not only must the table be clean and hygienic, but the kitchen table is a symbol of family interaction – single people don't need big kitchen tables – a place for family quality time which expresses family well-being. Moreover, the kitchen table is a site to cement intimate relations, for partners to spend time together. However, there are broader social relations at play over the kitchen table. Southerton (2001) argues that consumption practices that surround the kitchen table are different for different social groups. Middle-class kitchen tables are more likely to be regularly used for family meals and for dinner parties, while working-class kitchen tables are used less, on special occasions, and socialisation with friends is more likely to take place outside the home. It can be seen, then, that there is varying symbolic significance attached to kitchens and the type of social interactions that happen around and on the kitchen table, according to levels of economic, cultural and social capital. The symbolic dimensions of the kitchen and the kitchen table thus offer insights into individual, family and class relations.

Such analysis is important in cementing understanding of the ways in which consumption expresses social identities, symbolises class and status, and assists in delimiting cultural boundaries and networks in everyday practices and contexts. Central to achieving this is attention to the narratives and accounts that constitute the consumption act. Woodward (2003), for example, looks at consumption practices within Australian homes and the very differing emphasis placed on conspicuousness and style, on the one hand, and comfort and relaxation on the other.

Woodward argues that for studies of consumption and everyday life to be com-prehensive, the actors involved must be able to give their own account of, and reasoning for, their consumption and aesthetic choices. This is vital if the ways that consumption as a process by which meanings are continually managed through the accomplishment of a narrative, in conjunction with chosen elements of material culture, are to be uncovered. Thus, ways of living in the home, and the organisation and selection of the system of objects within its spaces, must be understood to be mediated by moral prescription with family, class and gender.

For example, it is clear that in broad terms the social construction of home is subject to intense attention in newspapers, on a plethora of television programmes, and in magazines. Images and representation of design, style and decoration within the domestic sphere all construct the home as a place to express individual taste, luxury, aesthetic expressions, luxury and management of identities. Conversely, however, the home is also often constructed as a place for relaxation, intimacy and emotion where style, design and conspicuousness are seen to matter little.

Woodward (2003) attempted to look into this dichotomy, and undertook research in different areas of suburban Brisbane. The study found that for middle-class women, aesthetic expertise and co-ordination are valued most highly, while in working-class homes order, cleanliness and family happiness are most valued. However, contrary to the popularity of a particular domestic visual style that is reflected in home decoration magazines, furniture stores and on television, in middle-class suburban homes there remains the enduring idea that the home should be a site for comfort and relaxation.

Even in middle-class suburbia – a place associated with family comfort and relaxation – consumption within the home to differing degrees resists the widespread processes of commodification and fetishisation of domestic decoration and aesthetics seen in the previous chapter. Related to this resistance is an emphasis on relaxation and a welcoming and comfortable ambience – in short, the ideal middle-class family home. So even though such preoccupations are more akin to working-class conceptions of home life and contrary to classical theories of consumption (such as Bourdieu's and Simmel's depictions of middle-class aspirations for distinction) there is a moral component which underpins the organisation and presentation of domestic space. This is underpinned by family relations, concerns for 'respectability' and a wish to create an ambience of domestic welcoming.

## Eating out: foodscapes and the city

Following on from the kitchen table, this section looks at what must be considered both as a fundamental part of our everyday lives and as having a profoundly important place in our urban landscapes. Bell and Valentine (1997) argue that food and restaurants have become important symbols of urban life. There can be little argument that eating out has become a cultural barometer, and the proliferation of eating places in the city has been one of the most visible and interesting features of contemporary urbanity. Family carveries, tea rooms, fast food takeaways, haute cuisine restaurants, food from around the world as well as local food specialities, gastropubs, street vendors, coffee houses and cafés are just a few of the foodie landmarks that can be found in our cities.

These culinary attractions have become more than just eating places, but rather are often developed as total consumption and lifestyle packages. They are not simply about just food and drink, but can be characterised in a range of ways that includes venues based on unfussy menus and home cooking, as being kid-friendly, or being about high fashion and cosmopolitan knowledge. Thus, it is clear that eating out is a container of many social and cultural practices, norms and codes (Bell and Valentine 1997).

For example, in several ways restaurants can be seen as theatres – there is an important element of performance that surrounds the nature of food service work (Finkelstein 1998). Performance can be seen as existing throughout the eating experience from the policy of selecting and recruiting waiting staff (often fun-loving, sociable and chatty) to the chef working backstage. And for customers, restaurants are places for gossip, political agitating, love matching, arguments and family celebrations. Crang (1996), for example, shows that the relationship between customer and server is very much a performative act. He describes the server as being 'on show' while doing mundane work, with banter between server and customer being an important element of the restaurant's experience. Moreover, with restaurants being one of the most fashionable places to be in cities, exclusivity and high fashion must be performed in trendy eating places. In a similar vein, the proliferation of food cultures from around the world is part of a middle-class distinction project for performing cosmopolitan knowledge. Bell and Valentine (1997) argue that patronising authentic and exotic restaurants in the urban restaurant districts means that you can 'travel around the world on your plate'. With whole parts of the city oriented and themed around food – you are where you eat. However, part of this proliferation is also a conflict with homogenisation of food, as in fast food takeaways. Hand in hand with contemporary trends for foodies to immerse themselves in the authenticity of different 'ethnic and exotic' cultures, which represents an interest in cosmopolitanism by the new middle classes, is the seeming continued growth of homogenised fast food outlets (see Bell 2002).

There are also important aspects of ritual and social relations surrounding urban foodscapes. With food outlets penetrating everywhere in urban areas – from theatres, art galleries, leisure centres and cinemas, to shopping malls, sports stadiums, airports, petrol stations and street corners – almost everywhere you go gives you the opportunity to eat. You can eat just about anywhere in the city but often with very different rules, and codes. For example, it is often forbidden to consume food not bought on the premises, it is certainly frowned upon to eat the previous customer's leftovers; eating too quickly or too slowly is also problematic. In restaurants of different types (and the myriad of other eateries that are found in our cities) there are clear structuring norms.

One of the most high-profile landmarks in the urban foodscape has been the development of supermarkets (see Case study 4.2 for a detailed analysis). However, Bell and Valentine (1997) suggest that outside the overdeveloped west, markets have a more central place in the consumer cultures of cities. Markets in many ways perform similar functions to those of the modern supermarket. For example, marketplaces are radiating centres, gatherings in which all aspects of life are shared, social life and communication take place, political and judicial activity are undertaken and cultural and religious events unfold. Markets are a site of exchange of news, information, gossip, social intercourse, where love and marriages are arranged, disputes, debts and arguments created and settled, and they are a place for drinking, dancing and fighting. Moreover, while for locals the market is the centre of social life, markets are also a magnet for visitors and tourists.

However, markets are not only a developing-world phenomenon; supermarkets have not yet managed to erase markets from western cityscapes. The special character of the market, revolving around lower prices, specialist stalls and the traders' patter, gives them continuing appeal. This charm indeed is so associated with fresh food and value for money that supermarkets often attempt to create a simulated market setting for their fresh produce sections.

## Case study 4.2 **The supermarket**

There are important urban consumption sites that have often been ignored by theorists in urban studies. Shopping at corner shops and supermarkets has, for example, been overlooked. However, the routine acts of weekly grocery shopping or nipping out for a pint of milk deserve close attention, and they are no less important urban cultural activities than flâneuring through the West Edmonton Mall or the Trafford Centre, Manchester.

From the point of view of the changing urban landscape, in fact, the supermarket is an incredibly important feature, since the policies of store location have a profound impact on cityscapes, with, for example, the trend towards large peri-urban sites leading to the re-routing of roads and dramatic changes in traffic flow, and associated spurts of growth in clusters on the urban fringe. Moreover, associated with the much talked about death and rebirth of city centres, supermarkets

**Figure 4.1** *Supermarket check-outs (Photo: Mark Jayne)*

have rushed to return to central locations. In sum, in terms of contemporary urban landscapes the impact of the supermarket is writ large.

Supermarkets are part of a most common and regular part of our shopping lives, and are thus seen as mundane compared to the consumption spectacles of mega-malls. However, the marketing strategies of supermarkets are becoming increasingly more sophisticated. The provision of exotic produce is central to these strategies, recoding supermarkets as places that provide for 'high' cultural pastimes – and cheap 'pile-them-high discount ranges' at the other end of the supermarket spectrum.

However, supermarkets are far from benign environments. Technologies of information gathering, self-scanning, singles nights, 24-hour shopping, loyalty cards and direct mailing campaigns have recently been added to other staple supermarket strategies such as muzak, crèches, bag packing, car loading, umbrellas and the piping of baking smells. However, supermarkets are not for everyone, and the urban foodscape is still also made up by enduring shops such as speciality shops and localised non-chain shops, and delicatessens. Nonetheless, supermarkets remain a big part of contemporary urban life.

Source: Bell and Valentine (1997)

## The second-hand city

One of the most startling contrasts that is present in the urban 'consumptionscape' is the relationship between the neatly stacked shelves of the supermarket or the neatly folded and ordered rationality of the high-street stores, and second-hand shopping. One of the most important areas of progress in studying the relationship between consumption and everyday life has been through research into second-hand shopping cultures. This interest represents a valuable contrast to studies of urban spaces such as shopping malls, boulevards, arcades and streets, and has focused on inconspicuous consumption spaces and a range of unconventional consumption practices.

This work has been at the forefront of unpacking the diverse nature of consumption in the city. Crewe (2000), for example, pays attention to how particular spaces are centrally implicated in processes of identity formation, and how consumers display complex, multiple and often contradictory consumption imperatives. Key to this understanding is that new areas have been deemed as acceptable to study – car boot sales, charity shops, retro-vintage clothes shops and discount shopping are now seen to be important topics to study. This represents a breakthrough in studies of urban consumption where questions of performance, discernment, distinction and spectacle, and issues of value and authenticity, are now firmly on the research agenda.

Encompassed in these studies of the second-hand city are sites and practices that range from jumble sales and familial and social networks (hand-me-downs) to a far more complex pattern encompassing multiple forms of second-hand exchange, characterised by a diverse set of social relations, from car boot sales to charity shops, retro shops, dress agencies, nearly-new sales, 'classifieds', auction clearances and antique fairs. Studies of such practices and sites have highlighted that skill and knowledge are important constituents of second-hand consumption cultures. Moreover, spaces of second-hand consumption constitute different spatial and consumption practices where encodings rest on intertwining understandings of second-hand in contrast to first-hand goods. What is clear, though, is that there is a diverse range of second-hand consumption spaces that are different in themselves. Thus, while second-hand spaces are about juxtaposition and combination, and generally thought to be used by those on limited income, they also offer distinction opportunities. For example, second-hand shopping offers the opportunity to extract maximum value and to secure bargains, as well as providing access to cheap retro-clothing often associated with the identity formation of different youth subcultures (Gregson and Crewe 2003).

Central to studies of second-hand cultures has been to show the progressive potential of this consumption activity. For example, Gregson and Crewe (2003) argue that those shopping for second hand are no dupes, nor heroes and heroines, but demonstrate their agency in different ways. Second-hand shopping is not just about constituting identity through goods, nor a materialisation of social relations. Instead, the ability of individuals to navigate the world of second-hand shopping is forged in geographical ways – people are required to look around and explore the nooks and crannies of their retail landscape. This means knowing the possibilities of places, and to be aware of where and when transient events like car boot sales and jumble sales occur. Other retail knowledge to be identified includes the best days and times for discounted goods at supermarkets and the purchasing of slightly out-of-date produce such as 'yesterday's bread' from market stalls. It is through such practices that sites are woven together as personal shopping geographies. Shoppers are required to spend much energy and to think hard – not just about what they might look at and/or what they may purchase, but also about what shopping itself is about in these spaces. Gregson and Crewe (2003) argue that people think hard about exchange, about value, about consumerism and use and need.

There are also other factors at play in the world of second-hand consumption. These are, for example, bound up with a number of issues such as the physical and symbolic properties of visions of second-hand exchange and how consumers constitute shopping geographies involving second-hand sites. These represent modes of being and relations of looking (and buying) that second-hand shopping reveals; and how these connect with a particular subject position. Gregson and Crewe (2003) show that there are core themes running through these interests around spatialities, underpinned by an oppositional understanding of the relations between first- and second-hand exchange. These are focused on a centre–margin metaphor including both symbolic distance and proximity, with first-hand exchange being loathed and revered, desired as well as shunned.

For example, car boot sales are located primarily in fields, car parks or open spaces on urban fringes and often held at 'inappropriate' times such as Sunday mornings (see Case study 4.3 for more on car boot sales). They are often seen as inappropriate by many local authorities who have attempted to regulate against 'shoddy goods'. In contrast, retro-clothes shops are often considered to be part of a landscape of gentrification. Charity shops, on the other hand, are very much aligned with standardised retail shops, and display on the high street. It is not just location that is important; there are different types of exchange taking place in different second-hand cultures. These include bartering and getting a bargain. Charity shops attempt to do retail properly, while retro-clothes shops tend to be wild and wacky with higgledy-piggledy display techniques and ambience. The practices of second-hand shopping are ones that are generally intended to be a radically different experience from first-hand shopping, and despite the various practices of second-hand shopping cultures, they do conjoin to suggest that shoppers have core premises associated with second-hand spaces.

Gregson and Crewe (2003) argue that there are general sets of assumptions held by the many different users of charity shops and, further, that within a diverse set of responses there were four common responses. First, and most significantly, was that second-hand spaces continue to be located by many within a moral economy that is both seen to be a local resource and grounded in need. These sites, then, are considered to be a resource not just for 'poor people' and people on restricted incomes, but a situated resource, where the traffic in things that occurs in second-hand sites is about redistribution within geographical areas. Sitting somewhat uncomfortably with this, at least in part, is a second premise: that the second-hand world constitutes a means of knowing place, being in place and encountering place. Reliant on the same identification between people, places and goods that lies behind the moral economy of redistribution arguments, the difficulty with this premise is that the second-hand economy provides low-cost goods not only for people who live economically restricted lives but for the more economically well-off also. For some this is a movement from knowing to unknowing places, while for others, knowing place through second-hand spaces can be akin to being confined in spaces that only allow access to goods that ultimately reinforce social exclusion.

A third premise is that second-hand spaces are enabling of multiple identities, and not identity in the singular. So, rather than encode particular visions of specific 'ideal' shoppers – as is the routine practice in many first-cycle retailers – these sites are widely seen to be ones that enable and juxtapose various forms of purchasing by diverse groups of shoppers. They are, then, sites that are regarded – at least ideally – as spaces that can accommodate those shopping from particular taste communities through to those looking for value, those shopping for fun, and those purchasing because they have to. Fourthly and finally, another core premise is that these spaces provide scope for resistant practices in one form or another. Encompassing both identity politics – for example, 'subcultural' style and partial resistance of the economically poor – and critiques of 'the system', second-hand sites continue to be widely regarded as distinctive from the high street and the mall or shopping centre.

## Case study 4.3 **Car boot sales**

The car boot sale is characterised by its lack of the fixtures and trappings of permanent forms of retailing, such as display stands, racks, mannequins, lights, music, and so forth. The car boot sale is reliant on altogether more impromptu domestic paraphernalia such as trestle tables, suitcases, tablecloths, plastic sheeting and cardboards boxes, not to mention the car itself (its boot, bonnet, doors and interior).

The presentation of goods 'for sale' at car boot sales reveals the extent to which this particular version of second-hand exchange is constituted from scratch. Its links with domestic consumption and its connections to other forms of second-hand exchange, most notably jumble sales, are clear. On the other hand, however, car boot sales are distinctly different from retro-clothes stores or charity shops where representational strategies are based on looking (and buying) in a similar manner to that adopted in other high street shops. Such characteristics and distinctions show how particular sites of second-hand exchange are associated with particular settings and practices of selling.

Participation in car boot sales is for both buyers and sellers and is centrally about establishing and reproducing principles of exchange. It is about working out how to buy, how to sell and how to constitute the transaction, and constituting a space of exchange that is different from retail spaces of first-hand goods. 'Naive, green, terrifying, novice' are some of the terms used by sellers and buyers to describe their opposite numbers. However, what is clear is that there are practical and accumulated knowledges that both buyers and sellers at car boots amass. There are a range of archetypal identities associated with car boot sales, from the middle-class women in 'good-quality' cars mobbed when first arriving at the sale due to perceived quality of products, to dabblers, collectors, commercial traders, and bargain hunters.

What is important to grasp about the 'alternative' consumption practices of car boot sales is the impromptu nature of encounters that characterise this version of exchange. Goods are being responded to as potential, not pre-planned, and the capturing value of 'getting a bargain' is more important than the goods themselves (which may not even be in working order). However, what is clear is that car boot sales are a site of second-hand exchange that requires participants to work out the principles of exchange that operate in and shape the sites, and how to work with them. And, particularly in its normative conventions, we can see how exchange is being constituted here as a game between buyers and sellers, as fun and as

continued

something that people take pleasure in. As such, there is more detailed coding and interpretation involved than in the more taken-for-granted first-hand retail world.

It is clear that car boot sales provide a theatrical and carnivalesque urban experience (see Chapter 5). The car boot sale is a stage on which people (if they wish) assume different personas through exchange, and a stage where the conventions of retail purchase normal in first-hand shopping are suspended. Car boot sales are a space where standard relations associated with the high street, the shopping centre, the mall, mail order or online shopping are subverted. Those who normally just buy are allowed to both buy and sell, and car boot sales are sites where buyers can contest the power of the seller to determine value, through money and through the exchange of knowledges too.

That car boot sales facilitate such alternative consumption practices and encode them in their very principles is a key factor in accounting for their popularity. Yet there is also considerable distance from the first-hand relations usually constituted around the two positions of exchange – 'retailer' and 'customer'. In the realms of first-hand exchange the retailer has the power to determine price and value for money. At car boot sales the possibility of 'finding a bargain' re-establishes the principle that the customer has the more powerful role in the transaction.

Source: Gregson and Crewe (2003)

## Old age in the city

What has become clear in this chapter is the increasing understanding of the role of consumption in the everyday construction of identity, lifestyle and forms of sociability. This has augmented the understanding of the importance of the role of ethnic and social groups such as lesbian and gay, ethnic and youth cultures in urban life. Such work has provided important visibility to often marginalised groups. However, Thorpe (1999) shows that even such studies of consumption have overlooked other marginalised spatialities and groups, those just 'getting on' and 'making do' in their everyday life. In particular, Thorpe identifies that people of post-retirement age, and people living in or using designated institutional space, have been ignored as urban dwellers.

Fokkema *et al.* (1996) show that despite an ageing population in western industrialised nations the proportion of older people living in cities is declining. Despite the growing political importance of 'grey power' and the increasing economic importance of the 'grey market' it would appear that consumption opportunities for older people in urban areas are highly restricted or often problematic. Issues such as the declining conviviality of local neighbourhoods, traffic congestion, street safety and fear of crime are important factors for older urban residents who might be less healthy and mobile. It is such factors that encourage 'grey flight' from the city to suburban, rural or seaside locations.

Nevertheless, the spaces occupied by older people in the city include multi-occupational residential housing, day centres and lunch clubs – but equally important is their experience of streets, shopping centres, tea rooms, parks, supermarkets, bingo and church. Thorpe's respondents were from working-class areas and the study focused on how older people constructed their selfhood and identity in consumer culture and what consumption means for this marginalised group. He found a diverse range of consumptive activities amongst older people from shopping strategies that take advantage of discounts targeted at old people, to low-cost meals at church and social clubs, as well as sociality associated with occupying public space by occupying benches in parks, streets and city-centre squares (see Figure 4.2). It is in this context that Thorpe introduces the issue of scale and also questions what resistance entails for older urban residents. He found that respondents' central concern was to be left alone, free to pursue mundane tasks such as buying alcohol or purchasing cigarettes, or spending their money how and when they wished.

Figure 4.2 *Older people in the city. (Photo: Mark Jayne)*

The age group that the research dealt with comprised those who had grown up with the optimism and promise of progress that characterised the modern period. The modern period with its promise of order, progress and individual and collective investment that would ensure the dream of the modern city had for many pensioners long since gone. They felt that they were isolated fragments in a world that has long since ceased to care for them – a crazy world, characterised by central and local government incompetence, blind greed and failure of people to do the 'right thing'. This representation and coding was also part of an ordering and production of space where they felt afraid and out of step with the rest of the world. Older people in the study felt left behind and marginalised by the pace of contemporary consumer culture, and despite the growing number of specialist products aimed at the grey market they remain a relatively invisible urban consumption subculture.

Thorpe thus looks at how the lived worlds of the pensioners were continually reproduced. He asked whether they are prisoners of space and whether the geographical experience of old age is one of constriction. He argues that the geographical lifespace of old age is characterised by an intensification of attachment to proximate environmental context and activities. However, it was not just that 'old age' meant that the respondents became necessarily (or totally) captives of space, but rather that they became captivated by 'new' spaces, 'new' socio-spatial productivities; and that space became an even more important element in thinking and placing the world. This suggests that physical, material, psychological and social constraints were key factors in the energising and developing of 'alternative cognitive maps' and social spaces which facilitated the re-routing of existential vitality – having to cope with adversity.

## Conclusion

This chapter has shown how consumption is at the heart of ordinary and mundane activities that make up our daily lives, and is an important constituent in the construction of urban spaces, activities, identities and social relations. We have seen how everyday activities are practices that can be knowing and transgressive, and have hidden codes and languages every bit as interesting and complex as spectacular spaces, places and consumption practices. Writers such as de Certeau and Lefebvre have identified the ways in which people engage with the spectacular urban landscapes in 'ordinary' and mundane ways and also how individual agency and resistance relate to consumption practices.

From negotiating, knowing and travelling through the city in our everyday urban worlds to inconspicuous consumption spaces such as car boot sales, charity shops, retro/second-hand clothes shops, markets, supermarkets and the home, studies of everyday consumption have been combined with interpretation of practices that constitute both the physical development of the city and identity formation.

Studies of everyday life have shown how consumption is an increasingly dominant aspect of social life. Writers have shown how ordinary people engage with that system, and that everyday activities are not simply determined or constrained by the dominant social order; on the contrary, everyday actions seek to undermine, deflect and short-circuit its power. Writers such as Lefebvre and de Certeau argue for a rich sense of the creative potential of practice. While this can give rise to a romanticised view of consumer resistance, in stressing the ordinary it is not about direct resistance (as in subcultures). Everyday life inevitably circumvents and subverts the dominant order from within. While people are largely powerless to challenge the system that is imposed on them, people escape without leaving, and in a consumer society the everyday activities of ordinary consumers exerts a promise of other ways of being and that consumers can actively produce their everyday lives – they make their own way in the world and even forge their own world through the everyday practices of walking, cooking, talking, reading, shopping, and so on.

---

**Learning outcomes**

- To have an understanding of the relationship between consumption and everyday life
- To be able to describe the ways in which ordinary and mundane consumption is a complex and rich topic for study
- To offer insights into how power and resistance are played out through everyday consumption
- To be able to critically appraise the relationship between everyday life and consumption in a range of different spaces and places

---

## Further reading

Hugh Mackay (ed.) (1997) *Consumption and Everyday Life*, London: Sage. An accessible text that reviews theories of consumption, and addresses topics such as everyday consumption and identity, broadcasting, and communication technology.

M. E. Gardiner (2000) *Critiques of Everyday Life*, London: Routledge. Offers detailed multi-disciplinary insights into the most important writers engaged in theorising lived experience and includes empirical research into everyday life.

Jukka Gronow and Alan Warde (eds) (2001) *Ordinary Consumption,* London: Routledge. This edited collection draws together some fascinating empirical research on the sociology of everyday life. The chapters included in the volume provide an important collection of case studies and theoretical ruminations.

Nicky Gregson and Louise Crewe (2003) *Second-Hand Cultures*, Oxford: Berg. A revealing exploration of the world of 'antiques', 'vintage', 'bargains', 'cast-offs' and second-hand goods. A useful review of the role of agency and resistance to first-hand consumer cultures.

# 5 Cities, consumption and identity

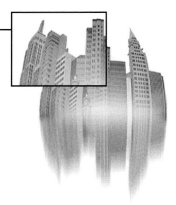

**Learning objectives**

- To look at the relationship between consumption and identity
- To explore the relationship between consumption and constructions of class, gender, ethnicity, sexuality and subcultures
- To think about how the relationship between power and resistance is played out through consumption
- To offer insights into how class, gender, ethnicity, sexuality and subcultural style are constructed differently at different times and in different places

While studies of urban consumption have predominantly focused on middle-class consumption cultures there has also been a growing focus on other urban identities. Working-class cultures, gender, ethnicity, sexuality and subcultural style are also central to urban consumer culture. This chapter reviews this progress, and looks at the experience of being poor in a consumer society and how the extremes of the 'haves' and the 'have-nots' are written on to the urban landscapes. Gendered consumption, ethnicity, sexuality and subcultural style are also discussed. The chapter is framed through Miles's (1998a) depiction of the 'consuming paradox', which shows that while individuals feel they can construct their own identities and sovereignty through consumption, consumption simultaneously plays an ideological role in controlling the character of everyday life – within a rationalised social framework which structures, enables and constrains our urban experiences.

The opportunity to consume is a key indicator of power structures and is a key part of how social hierarchies, relations and everyday practices and processes are constructed. Moreover, consumption can be understood as playing a key role in codifying power as social space in the city. In 'contested' spaces in the city, identity is constructed and regulated, and the social construction of difference, negotiation of identities and power struggles takes place. However, urban spaces are often inhabited by many different social groups and thus have multiple meanings for the many different people who mingle in those urban spaces. It is in these terms that the codifying and interpretation of the city through studies of consumption involve knowledge of actual usage, power relations and discourses in everyday life (Shields 1992c; Massey 1993; Ryan and Fitzpatrick 1996; Pred 1996).

When we understand that social space is codified in geographic and socially specific ways, then it is clear that a variety of social identities are produced and reproduced through consumption in those spaces. Social identities are not fixed but are dynamic, and cleavages that work along various separate dimensions of social stratification (for example, the intersections of class, gender, ethnicity and sexuality that people experience) are constantly being produced and reproduced in different places and at different times (Keith and Pile 1993). In order to explain how this is played out in urban spaces and places there is a need to explore the multi-constituencies of political, economic, social and cultural factors in place and time (see Warde 1997). This argument highlights the importance of theorising how consumption constitutes the experiences not only of those who have ready access to the cultural and economic capital of consumption, but also of sectors of the population who are marginalised or disenfranchised from consumer culture. Such a viewpoint acknowledges that consumption is necessarily an area of conflict – as individuals, social groups, cities, regions and nations try to come to terms with all sorts of stresses and tensions (see Case study 5.1).

## Case study 5.1 Consumption, identity and conflict: consumerism in South-East Asia

Over the past thirty years economic growth in many part of South-East Asia has been dramatic in countries such as Indonesia, Korea and the city-state of Singapore. For example, Chua (1998) discusses the development of consumerism in Singapore, which has been bound up with the city-state's rapid economic development and success in securing itself a place in the hierarchy of global cities. During this time, economic growth has almost eliminated abject poverty, and in the 1990 national census 'window shopping' was ranked the number one leisure activity conducted away from the home. Expensive cars dominate the city's streets and almost every international fashion house is represented in the city, from mass market to exclusive couture (such as Issey Miyaki, Sonia Rykiel, Donna Karen and Gianni Versace) as well as globally significant shop and restaurant chains. Overall, household expenditure on consumption activities has increased dramatically – for example, eating out in restaurants or from other vendors as part of total food expenditure per household increased by more than 10 per cent between 1977 and 1987.

However, the emergence of this consumer culture has been received uneasily by the national government and also to a certain degree at a popular level. Such disquiet has been codified locally as an ideological and moral discourse with two explicit dimensions. First, there is a concern that consumer culture represents an 'excess of affluence', and with the deprivation prior to industrialisation fresh in the minds of the people of Singapore, the function of utility over fashionability is still valued. Secondly, there is a critique that consumer culture is overly dominated by 'westernised' or 'Americanised' commodities and cultural production. Key to this perspective is that Singapore is non-western and must strive to maintain its distinctiveness.

Elsewhere in South-East Asia concern for the maintenance of other national identities also ensures wariness of the spread of consumer culture and what are seen as practices that are destructive to traditional values. For example, worries have been expressed in Korea that westernisation will lead to an erosion of Confucian ethics. In Malaysia and Indonesia concerns that credit expansion and conspicuous consumption will aggravate income inequalities have been prominent.

As part of Singapore becoming inundated with American-produced consumer goods and entertainment, policies that seek to protect national sentiment and identities have been initiated. However, such anti-western and anti-American concerns would appear to be contradictory, given that the city-state relies heavily on Japanese goods but no fear of Japanisation of Singaporean culture is expressed. The difference lies in the 'cultural' component of goods from the two countries. Japanese goods tend to be branded as quality, technology-based goods, such as cars, televisions, fridges, radios, computers, and so on. On the other hand, the cultural and ideological content of goods and services (such as mass entertainment) is seen as a harbinger of western/American 'liberal values', moral laxity and individualism, emanating from countries where individual rights and self-interest lead to high divorce rates, crime, promiscuity and drugs. There is a fear that such western liberal individualisation will undermine local traditions and social cohesion and threaten 'Asian values'.

This concern is particularly focused around the attitudes and behaviour of young people who are seen as most at risk from the flows of western cultural products.

continued

The conspicuous consumption of designer and brand names, western music, food, and often shocking fashion represents a discourse of excesses that is seen to undermine traditional values of frugality and a past that is fondly remembered and appreciated. Such discourses have also been augmented by a growing confidence in Asian consumer products. Giordano, for example, a successful Hong Kong company that manufactures youth casual wear, has been successfully competing with western brands, and a new generation of adventurous Asian youths are developing their own consumer and cultural products.

One of the most important outcomes of such popular and political discourses surrounding the spread of western consumer culture has been moves to curb excessive consumption. This has been focused, for example, around institutional control over credit, and since 1992 a qualifying annual income for obtaining credit cards has been in place. Moreover, each individual's credit limit is set so as not to exceed twice their monthly salary. Loans on mortgages and cars are restricted. While such close legislative control on spending led to a downturn in the retail sector in Singapore during 1995, leading to significant financial losses for large stores, downsizing and outright closures, the very high standard of living enjoyed by many of Singapore's citizens has offset the economic problems and social dissatisfaction that such control might have caused.

Source: Chua (1998)

The multidimensional nature of the consuming experience, and the extent to which consumption is central to the construction of the relationship between, for example, individuals and society, clearly come to light (Wynne and O'Connor 1998). Sections in this chapter examine this contention in terms of the ways in which class, gender, ethnicity, sexuality and subcultural styles are central to urban consumer culture.

## Class

Chapters 2 and 3 showed how the struggle for bourgeois control over working-class consumption cultures has been a defining feature of the development of the modern and postmodern city. Crompton (1996) suggests that urban theorists have generally sought to conceptualise this relationship between consumption-related trends and class analysis in two ways: first, in the linkages between consumption practices and class formation; and secondly, in questioning the extent to which a shift to the service sector and consumption-related employment has transformed employment relations and the experience of work. The latter represents the postmodern perspective outlined in previous chapters and focuses on the consequences of the breakdown of 'class' identity and consciousness.

What these two distinct approaches have in common is the view that social classes are in part constituted through cultural practices, including patterns of consumption. However, while such understandings have been present throughout theoretical engagement with urban consumption in the work of Veblen, Simmel and Bourdieu, the most significant weakness in their work was a failure to engage with social identities beyond middle-class consumption cultures. Sociologist Zygmunt Bauman has provided an important antidote to this weakness by looking at the relationship between poverty and consumption.

Bauman (1998) shows that in recent years, at times of economic instability, politicians of all parties speak in unison about a 'consumer-led recovery'. He argues that this rhetoric is uttered in the context of falling manufacturing outputs, empty order books and sluggish high-street trade. These problems, he suggests, all tend to be blamed on a lack of consumer confidence (which means that consumers' desire to buy on credit is not strong enough to outweigh their fear of insolvency). Bauman suggests that, ultimately, political hopes of overcoming all these structural economic problems are based on consumers 'doing their duty' – people buying, buying a lot and buying even more. In these terms, economic growth is the key measure of things being normal – in good order – and is in a large part based on the spending of consumers. Bauman notes that this current situation supersedes the historic central political concern of the industrial period that was based on the productive strength of nations – that is, a healthy and plentiful labour force, full coffers and daring entrepreneurship of capital owners and managers. In other words, the role once performed by employment in linking together individual motives, social integration and economic activity has now been assigned to consumer spending.

It is in this climate that Bauman (1998) reflects on the construction and experience of being poor in a consumer society. He suggests that the phenomenon of poverty does not boil down solely to material deprivation and bodily distress. He argues that there are social and psychological conditions ensuring that we judge our position in the world, and underpin our wish to fulfil the standards of decent life that are generated and practised by any society. He argues that inability to abide by such standards can cause distress, and concludes that poverty means being excluded from whatever passes as normal life. In a consumer society, then, 'normal' life is a life filled by consumption, 'people preoccupied with making their choices among the plethora of publicly displayed opportunities for pleasurable sensations and lively experience' (Bauman 1998: 37). Bauman also shows that boredom is a common complaint of the poor, and in a consumer society a happy life is defined as being insured against boredom, with something new and exciting always available to purchase. In a consumer society the poor are socially defined as blemished, defective and deficient – in other words, inadequate consumers. However, Bauman importantly highlights that the poor do not inhabit a separate culture from the rich and thus poverty is aggravated by economic growth, just as it is intensified by recession, so when the economy grows, the poor get poorer and the rich get still richer (see Case study 5.2).

## Case study 5.2  **Discount shopping: from Woolworths to Wal-Mart**

Shopping is one of the few activities that bring all classes of people together in a public (or somewhat public) place. Sharon Zukin suggests that one particularly important site for social mixing oriented around consumption is discount stores. Zukin describes discount stores as contemporary public squares with the social functions of marketplaces. They are locations where everyone can surround themselves with shiny, tawdry, giant-sized commodities and standardised products – in discount stores everyone can find a bargain. Zukin provides a genealogy of discount stores and identities two particularly important discount chains, Woolworths and Wal-Mart.

Zukin describes the growth of Woolworths from the 1870s, when all the goods in the self-service stores had the same low price of five cents – hence the phrase five-and-dime (a term that is still often used to describe discount stores in the US). Woolworths invited everyone to be shoppers (unlike department stores), and the merchandise lines sought to appeal to everyone. Five-and-dimes attempted to attract shoppers by romanticising variety and offering a sensual shopping environment. Woolworths had very distinctive red and gold facades, and stacked small items in eye-catching pyramids that could be viewed through plate-glass windows. The self-service ethos of Woolworths allowed young women to handle sensational products, such as red lipsticks, nail polish, hair clips and costume jewellery, and children could test yo-yos and whistles, and handle embroidery kits and toy carpentry tools. Luncheon counters were introduced into Woolworths in 1907, providing fast food convenience for office workers and a place for people to meet and socialise over a coffee or coke.

During the 1970s, Woolworths shut its flagship stores on Fifth Avenue in midtown Manhattan, and in 1997 closed all its remaining five-and-dime shops due to growing competition in the lucrative discount stores market. For example, by the

1960s, discount stores such as Wal-Mart, Kmart and Target had ushered in a new order of shopping, underpinned by much larger stores and the scale of product ranges on offer to the consumer. Sam Walton, who started Wal-Mart in 1962, sought to generate a folksy mythology that Wal-Mart was a style discount store and a space to find a real sense of local 'community' while finding a bargain. Of course, in order to deliver this, Wal-Mart had to generate a global system of purchasing and distribution that could stock the huge stores and to profit from economies of scale. Wal-Mart was also particularly important in that it located stores in areas and regions of America that had been a shopping wilderness, arguing that it was allowing local residents to fully join the worldwide consumer revolution for the first time. Utilising brand names as loss leaders, paying low wages and locating stores in low-rent areas, Wal-Mart has seen an incredible growth in its fortunes, and the company is currently the second largest employer in the US after the federal government.

Sharon Zukin (2004) describes the arrival of Kmart (a subsidiary of Wal-Mart) in the trendy New York district of Greenwich Village during the late 1990s, with its racks of cheap clothes and plastic storage bins and its meagre grocery department specialising in snacks and convenience food. However, despite its discount-store status, Zukin suggests that 'everyone' – no matter what their social background – goes to Kmart to shop. Nevertheless, despite the popularity of Wal-Mart stores, opposition that began in small towns has spread to cities such as New York amid concern that communities and neighbourhoods can be ruined by the blank outer walls of the stores, absentee corporate owners and heavy traffic. The debate over such superstores is partly aesthetic and also about a sense of place; these stores are ugly, turn their back on the street, and buffer themselves with acres of parking space. Their economies of scale also often put local merchants out of business.

Despite such opposition to Wal-Mart stores, or similar companies such as Costco, Barnes and Noble, Target and Home Depot, by 2000 these major discount chains were amongst the top ten retailers in the USA, and Wal-Mart has been buying up chain superstores elsewhere in the world. Zukin concludes that the seemingly unabated growth in discount stores, while being based upon shopping for all, is also largely due to poorer shoppers who overuse their credit cards. She concludes that Wal-Mart does not want to eliminate the importance of social class in its discount consumer space; it just wants us all to keep shopping.

Source: Zukin (2004)

In order to explain this situation it is useful to turn to Beverley Skeggs's (1997) study of working-class women in the north-west of England. Skeggs identified how 'respectability' was always an issue in the subjective construction of working-class women's identities – as people sought to 'get by' in a world where middle-class values dominated. However, Skeggs shows how 'working class' is a category that came into effect through middle-class conceptualisations of social standing. 'Working class' is thus a method of distancing the middle classes from those they consider 'beneath' them; making the 'other' knowable, measurable and organisable. This shows how class is a discursive, hierarchical, specific construction and a product of middle-class political and social consolidation. Importantly, Skeggs shows how this is underpinned by a daily process of reiteration. This means that 'categories of class operate not only as an organising principle which enables access to and limitations of social movements and liberation, but are reproduced at an intimate level of everyday life, and through practices such as consumption (Skeggs 1997: 6).

Skeggs draws on the work of Bourdieu, who thought that class was neither an essence nor an intermediate set of fluctuating signifiers, but an arbitrarily imposed definition with real social effects. He outlined types of 'capital' as being key to constructions of class in three ways. First, Bourdieu showed the importance of economic capital to class constructions. This includes income, inheritance and monetary assets. Secondly, Bourdieu identified the importance of cultural capital to middle-class identity. He described cultural capital as being made up of three forms: as an embodied state, that is, in the form of long-lasting dispositions of the mind and body; in an objectified state, in the form of cultural goods; and in an institutional state, resulting in such things as educational qualifications. Finally, Bourdieu noted the importance of social capital to constructions of class. He showed that social capital is resources-based, is part of what connects people and forges grouping membership, and is generated through relationships.

Skeggs argues that legitimisation of these forms of capital is the key mechanism in the conversion of power. Thus, people are distributed in the overall social space according to the global volume of capital they possess, the composition of their capital and the relative weighting of their overall capital. As such, the social space we occupy has been historically generated. For example, this explains 'why those with a small volume of cultural capital will have difficulties increasing its composition and will subsequently have a circumscribed trajectory' – resulting in an inability to escape the restrictions of their class (Skeggs 1997: 26). Working-class people can carry markers of their poverty on their bodies, as illness or radiating lack of confidence, deprivation or aggression. Class is lived culture that can be mapped on to the city, and working-class life is characterised by 'making do' and 'getting by'. Skeggs asserts a mode of class which is based on 'capital movements' through social space – and, as we have seen, it is bourgeois wealth and sensibilities that dominate urban space. In other words, the physical structure and our everyday experience of space are determined by the distribution of various forms of capital. The physical city and urban life have been historically generated from previous movements of capital and are manifest in different opportunities in the labour market and the education system. Similarly, consumption is part of everyday and structural divisions which organise what everyday experience and opportunities are available. The working class are continually trying to halt losses rather than trading up and accruing extra value – and simply reproduce the very distinctions they hope to transcend.

Skeggs (1997) explains this in simple terms by showing that the discursive construction of class is underpinned in structural and everyday ways. This shows that ways of walking, talking and dressing make it difficult for working-class people to 'pass' – in middle-class company, or to gain professional, managerial or professional jobs. However, the need of the middle classes to distance themselves from working-class life led to the 'corralling' of working-class consumption practices into enclaves in the city, as well as to working-class culture being banished from bourgeois-dominated urban consumption spaces, such as arcades, department stories, shopping malls, urban villages and city centres (see Case study 5.3). This conversely explains why working-class consumption practices such as bingo, dog racing, darts and some clothes, music and food have not been colonised by the middle classes.

## Case study 5.3 **Working-class consumption, carnival and urban cultures**

The attempt to dominate the city by urban bourgeois elites is a key characteristic of the development of the modern and postmodern city. These struggles have ultimately been founded on the expulsion of working-class people into particular residential and industrial areas and, further, through the banishment of working-class people and consumption from urban spaces and places. One area where this conflict was particularly visible throughout Europe was in battles over 'carnivalesque' leisure and consumption activities. This battle was not simply about leisure activities, however, but was at the heart of political, economic and social urban change. Struggles over the carnival represent broader class conflict over everyday structures of work, lifestyles and leisure, and show how bourgeois values and agendas were imposed on our cities (Stallybrass and White 1986).

continued

Loosely based around religious tradition, carnivals and fairs were a prolonged and subversive escape from the everyday drudgery of pre-industrial and indus-trial working-class life. Full-scale drunken brawls were commonplace at carnivals, as well as 'other pastimes guaranteed to satisfy the basest of instincts of our predecessors' (Edwards 1996: 7). As industrialisation and urbanisation went hand in hand, so the carnivals and fairs became larger and, it is said, more hedonistic. In sum, the atmosphere was 'heady and intoxicating', and while these events were initially classless occasions enjoyed by industrialists, workers and country gentle-men, class tensions became progressively more prominent.

Throughout the eighteenth century, opposition to the carnivals and fairs grew, not only within the Church, but also amongst the emerging bourgeoisie. Crimes such as pickpocketing, alcohol-induced violence and the pursuit of blood sports were seized upon by various interest groups who wished to undermine (or better still to do away with) carnivals. As Edwards suggests, while some members of the middle classes enjoyed the festivities as much as 'lesser mortals', others begrudged the artisans their annual 'knees-up' (1996: 11).

Nevertheless, even in the first half of the eighteenth century many 'lords' or 'gentlemen' were not entirely disengaged from the culture of the common people, being involved in cock-fighting, fishing and coursing. There was still considerable co-dependence between landowners and agricultural workers, and between the industrial overlords and their workers. As such, '[t]he dominant attitude during the first half of the eighteenth century reflected a subtle blend of tolerance, self interest, and the paternalistic habit' (Malcolmson 1973: 70). However, although gentlemen were still involved in such recreations, they were culturally split by the pull of the 'sophisticated city', which looked down on rural culture. By the end of the eighteenth century there emerged a distinction between those who continued to accept the traditional customs of the people (and the more traditional landed gentry), and those who regarded them as anachronistic and unacceptable popular impositions (the urban bourgeoisie).

In sum, increasing disdain at this time for the unsophisticated traditional rituals and carnivalesque social forms was fuelled by the (initially gradual) development of a small bourgeois class emerging out of industrialisation. This class was largely educated in Nonconformist schools that instilled the values associated with the Protestant work ethic. Such schools bred a concern for labour discipline and effectiveness, combined with the religious values of frugality, temperance and industriousness. At this time, carnival was being pushed to the margins of the landscape (see Shields 1991). For example, carnivals and fairs, which had historically been held in urban centres, were gradually moved to more peripheral sites, often because of pressure from traders, who objected that the fair had all the attributes of the marketplace, and did not have the 'proper' behaviour seen at the more 'proper' department stores and arcades.

Across Europe it became increasingly difficult, with the development of the 'economic' as a separate conceptual sphere in the unfolding of bourgeois thought, to countenance the muddling together of work and pleasure/leisure as they regularly occurred at the fair (Stallybrass and White 1986: 30). The middle class systematically disengaged from carnival practices – and promoted 'rational recreations' such as sport and woodwork for the working classes (Malcolmson 1973). The carnival thus became central to the bourgeois imagination in defining everything it was not.

However, despite bourgeois success in dominating urban life, such conflicts continue to play an important role in the contemporary city. For example, a 'moral panic' has grown up in the UK around city-centre drinking cultures. In the past decade urban regeneration initiatives have developed night-time economies, and alongside traditional pubs new hybrid café/bar/club venues have emerged. However, these consumption landscapes have promoted concerns about alcohol-fuelled disorder, drunken brawling, public sex acts, and litter from 'takeaway' food wrapping strewn across streets, which are at odds with a cosmopolitan-framed urban renaissance (see Chapter 7). The UK Government's White Paper *Time for Reform* (2001) has given the local state new powers to tackle this by shaping local landscapes of consumption. Once again, working-class consumption cultures are seen as a threat to urban bourgeois sensibilities.

**Figure 5.1** *Regulation aimed at tackling drunkenness in public space. (Courtesy of Paul Chatterton)*
*Source: Chatterton and Hollands (2003)*

# Gender

At a cursory glance the urban landscape is unlikely to reveal obvious gender differences in the same way that it is easy to see expressions of class and ethnicity in our cities. Men and women inhabit the same areas; however, despite this co-presence, men and women not only use cityscapes in different ways but also experience and perceive them differently (Valentine 1989). Because gender roles and relationships are largely taken for granted, the gendered nature of the urban experience was also ignored until relatively recently. So too was the extent to which cities and their spaces and places reinforce gender differences and inequalities.

Nixon (1996) suggests that recent years have witnessed a steadily rising interest in men's relationship to consumption through studies of designer menswear, cosmetics and glossy style magazines. However, while studies of urban consumption have stressed diverse codes amongst men – presentation of self in terms of the cut of suits, the purchasing of trainers, cars, hi-fis, holidays and drinking – the way in which women have been involved in the economic, social and cultural changes associated with the city has been of more profound importance. Domosh (1996) argues that the fact that women constitute the major class of consumers in modern western society is incontestable in everyday life. Although men participate in retail activity, both as owners and workers in shops, and as consumers, women's involvement in retail is more complex and has had a more meaningful impact on the development of the city. Consumption is a defining characteristic of femininity and has played an important role in women's social and economic position and relationship with the city.

One of the key ways in which this has been constituted was seen in Chapter 2, which highlighted the distinctions between the private and public that make up the urban realm. The public was linked to the city and working men, whereas the private was linked to the suburbs and women. Underpinning this distinction was the recognition that men and women traditionally performed very different roles in society – with men travelling to work each day and women remaining at home. This represented a gendered division of labour that had spatial implications. The suburbs were the 'natural' space for women and the city for working men. In these terms the city could be divided (at the level of ideology) into discrete male and female zones.

This point is pursued by Glennie and Thrift (1996), who look at the way consumerism in retail environments has historically been gendered. They review the history of retailing power and bourgeois urban dominance through the separate spheres of the male-dominated public/political world, and a female domestic/private realm (although the home sphere is dominated by masculine power).

Glennie and Thrift argue that consumer culture in the modern city underpins this public–private divide. For example, men's consumption was important, noteworthy, skilled and rational; women's was trivial, routine, unskilled, unreflective or irrational. Women were positioned as wives, mothers and daughters who looked after the home, a nest for the male breadwinner. Once established, these gender discourses strongly reinforced highly gendered social relations. In short, women's consumption activities were initially restricted to servicing the household and to maintenance of social standing (Stevenson 2003).

However, such denials of women's competence in the public space of the early modern city posed problems for retailers, whose numbers were increasing in the accelerating commodification of everyday life. It is in this context that the department store has received much attention. Department stores are seen as being constructed around the gendering of consumption. The department store brought order to chaos, making part of the 'public' safe for women. The department store taught women how to be good consumers, and made encounters with goods a norm of everyday life. This was achieved through design and control of (gendered) space, adverts and display of commodities. Shopping became a skilled knowledge-based activity, with consumers' knowledge controlled by retailers and advertisers. Glennie and Thrift (1996) suggest that, when part of the shopping crowd, women avoided threats to their respectable middle-class identities. The dangers of possible encounters with men, the working-class and ethnically diverse strangers were off-set by the safety of numbers offered in crowded streets.

In department stores women faced an incredible array of goods, and recreated and asserted women's role as shopper and thereby ascribed values of femininity to shopping. Domosh (1996) shows how consumption spaces of the city were fundamentally shaped by the relationship between gender ideology and the development of modern consumer culture. Femininity and consumption were defined in relation to one another. For nineteenth-century women it was not the home that was a refuge from men, but the city streets that offered a break from domesticity and became an arena for action. This represented an inversion of the home/private dualism, as women – and, in particular middle-class women – increasingly played an important role in shaping city life, albeit in a process that they had little control over.

However, women's participation in public life ultimately helped to maintain the existing social order. Women served the family through purchasing commodities, and department stores played an important role in consumerism catering for women. It is in these terms that Domosh describes feminised retailing districts in late nineteenth-century New York. These were characterised by ornamental architecture and grand boulevards, restaurants, bars, small boutiques and large department stores. This was an urban landscape designed for consumption and it was women who were the main patrons of its shops. In these shops, qualities associated with nineteenth-century femininity and the domestic sphere came together in well-ordered displays in a safe environment. Central halls were like cathedrals, and shopping became a moral act, a religious duty for women. Thus, the feminisation of consumption rendered women's presence in central urban areas as acceptable.

Jackson (1997) argues that this domestication of the city had further dramatic effects on the cultural landscape of the city. Department stores looked like parlours, with tea rooms, restaurants and art galleries. Outside the stores, wide galleries and sidewalks allowed fashionable dresses to stay clean and, increasingly, gas and electricity added to the propriety of the area for women. Rapid transit systems gave middle-class women easy access to residential enclaves, and the growing presence of women within central areas of the city acted to further legitimise the wealth and power of the bourgeois – further underpinning the relation between gender, ideology and class and the development of the consumer society. For example, Gail Reekie (1992) looks at the sexual transformation of a Brisbane department store in the 1930s–1940s and alerts us to some of the sexual meanings and processes of gender formation implicit in urban and public spaces, and to the strong association between consumption and femininity (see Table 5.1). She also shows that in the 1930s it was considered unmanly to shop for goods other than men's own clothing and products associated with masculine activities and interests – such as electrical and sporting goods and tobacco.

Commercial districts thus provided meeting places, such as tea rooms, for women who would not patronise saloons (Zukin 1998a). The expansion of retail trade created many new jobs for women and places perceived as safe for women shoppers and children. But these opportunities presented women with special problems. Like the wages of actresses and nightclub performers, sales clerks' salaries were low. These occupations often recruited young, unskilled, attractive women who had migrated to the city, and, faced with poverty, some of those women despaired and either tried to marry or turned to prostitution (Zukin 1998a). The multiple identity positions that are bound up with gender, class, ethnicity, age, and so on ensured that individual women's experiences of the burgeoning consumer society were thus very different.

However, Stevenson (2003) suggests that debates over the relationship between urban change and gender have often been messy and placed too much emphasis on gendered stereotypes of consumption. This contention is based around two key aspects. First is the idea that there is a gendering of objects into sex-specific commodities, from fast executive cars and power tools to jewellery and cosmetics.

**Table 5.1** *Organisation of a Brisbane department store in the early 1930s*

|  | Female | Female/Male | Male |
|---|---|---|---|
| 3rd Floor | Crockery<br>Cake shop<br>Lending library | Furniture<br>Electrical<br>Hardware<br>(including kitchenware<br>and café) |  |
| 2nd Floor | Curtains & soft<br>  furnishings<br>Linos & carpets<br>Toys<br>Grocery | Sporting goods<br>Chiropodist<br>Dentist |  |
| 1st Floor | Women's clothing<br>Underwear & corsets<br>Millinery<br>Maids' wear<br>Babywear<br>Beachwear<br>Hairdresser |  | Travel and goods<br>Men's & boys' clothing<br>Hats<br>Tobacco |
| Basement | Jewellery<br>Stationery<br>Hosiery<br>Haberdashery<br>Gloves<br>Ribbons & laces<br>Needlework<br>Handbags<br>Fancy goods<br>Cosmetics<br>Confectionery<br>Toiletries<br>Canvas goods<br>Dress material | Shoes (separate men's<br>and women's sections) |  |

Source: Reekie (1992: 178)

Stevenson shows that this neglects the understanding that there is nothing intrinsically masculine or feminine about such products, but rather that they are invested with gendered meanings and associations (the only truly gendered commodities being tampons and other sex-specific or medical products). Second is the argument that a gendering of consumption practices has taken place. For example, it is said that men do not like shopping, and women do – and more specifically that women go to supermarkets and department stores and men go to do-it-yourself stores, and buy CDs and electronic equipment. Stevenson argues for a more critical approach in order to better understand the gender ideologies that are bound up in consumption.

It is in this context that Lury (1996) is right to argue that the position of women in relation to consumption is contradictory. Women themselves are seen as active participants within the constraints of their positions, but this is dependent on class, ethnicity, sexuality and life course. For example, Lury argues that although young, affluent and independent women with inherited good looks may enjoy a sense of power and self-worth, older, poorer, economically dependent or less attractive women may feel excluded or derided. Feminists have thus interestingly explored how urban policy and the structuring of cities, as well as urban studies, have marginalised or ignored the needs and priorities of women.

Moreover, there is also a sense that the safety of urban consumption for women has been romanticised, and that in their everyday lives women have to deal with a range of invasions of their privacy in the urban crowd, from personal comments and inappropriate gestures to more serious incidents of physical violence. Femininity therefore brings with it a range of restrictions in the use of public space. These restrictions may arise from actual experience of urban violence or from women's fear of violence (Valentine 1989). Research has shown that many women, especially older women, are afraid to use certain spaces such as city streets and parks, particularly in the evening or at night, because of fear of attack, especially by gangs of youths. Feminist critics have suggested that one of the reasons for women's relatively restricted access to a range of public spaces is the general assumption that women are in need of protection from the hurly-burly of the public arena. Women's construction as dependent on men, both economically and morally, or as lesser beings – as fragile and in need of protection – reduces their rights to freedom in public space (Allen *et al.* 1999). It is clear that the gendering of urban space is an important element in the constitution of social identity and that the spatial locating of women is used to construct that identity in certain ways.

Conversely, it is important to stress that the growth of cities has brought greater freedoms for women. City living was and is important in the development of new forms of association between men and women. For example, Wilson (1992) argues that the anonymity and excitement of the late nineteenth- and twentieth-century city were crucial in the rise of feminist politics, as urban living often brought freedom from patriarchal control. The 'new woman' of the twentieth century challenged the sanctity of hearth and home, and the city brought welcome anonymity and new opportunities for public association with a wide variety of women from different backgrounds. Feminist historians have documented in detail the lives and struggles of the early pioneers of women's education and employment, women involved in the suffrage movement, lesbians, and women who searched for new, independent living in the city (see Wilson 1992).

What is clear, then, is that the relationship between gender, consumption and urban life is intertwined – each informing, mediating and intimately affecting the (physical, political, economic, social and cultural) changes associated with the modern city. Consumption was, and remains, a defining feature in notions of masculinity and femininity and notions of the public and private realms. Women negotiated and imposed their position within the urban landscape, albeit within the restive prescriptions of the patriarchal and capitalist city life. However, while it is clear that gender and consumption are defined in terms of each other, we must not ignore the fact that women have differential resources and opportunities to participate in urban consumer culture.

## Ethnicity

Sharon Zukin (1998a) suggests that mixing of things and people that were culturally strange was a key feature of the early modern city and that the development of consumer culture was underpinned by central commercial districts of the city that were ethnically diverse. Urban growth was played out against a backdrop of social and demographic change of mass migration and the rise of entrepreneurs from ethnic minorities. It was at this time that different types of marginalities, including immigrant, ethnic and commercial cultures, combined in new ways to produce urban consumption cultures. In the early modern city, then, for a brief time, both spectacle and tolerance were common cultural denominators (Zukin 1998a). This was seen in restaurants, cafés, department stores, hotels and shops. Visible marginal groups contributed to urban cultures that brought outsiders inside.

However, Zukin suggests that this diversity was short-lived and that the bourgeois elite's success in rationalising working-class work and leisure time into strict spatial areas led to increased social segregation around racial lines. Moreover, the 1930s depression, economic hardship and social unrest also contributed to the city becoming more socially and spatially segregated. As we saw in previous chapters, this urban segregation was most starkly seen in the growth of ghettos. However, despite the growth of highly visible consumer culture in commercial cores, the ghettos and other working-class areas remained largely untouched by commercial investors. Ignored by major department stores, big chain stores and retailers selling high-quality goods, urban ghettos have only recently begun to attract the attention of corporate planners.

Perhaps the more obvious link between ethnicity and the development of urban consumer culture is shopping streets frequented by immigrant and native-born minorities. Such streets have had a vital impact on urban and ethnic identities in cities throughout the world, where shoppers, peddlers, managers and clerks are likely to be Africans, Koreans, African-Americans, Indians, Kosovans, and so on. In these areas of the city, shop fronts and signs are in many languages. Newspapers and foods from around the world are commonplace. This represents not the aestheticised cosmopolitan commodity worlds of gentrifiers but a transnational consumerism that has created particular urban identities and lifestyles. These interactions are a cultural juxtaposition in spaces of consumption and indicate a 'hybrid' urban culture (Bhabha 1994) not necessarily dominated by corporations or the middle class – on the street, diversity can thrive.

One of the most visible of these, for example, is the shopping streets created by new African-American identities. Some writers suggest that such diverse consumption cultures are a key indicator that consumption is not a passive act but represents the richness of cultural struggle in and around identity. For example, Paul Gilroy (1987) suggests that the night-time is a time when black consumption is prominent and there is a refusal and reversal of the dominant white tendency to privatise consumption. In black culture, consumption is celebrated as a collective, affirmative practice in which an alternative public sphere is brought into being (also see Back 1996). He gives the example of music consumption, and argues that records are used as a cultural resource in processes of creative improvisation in response to the requirements of public occasions – religious, political and cultural. In the black dance hall and at carnival, music played by sound systems is racially coded to produce a public association with black people that is not created through racist stereotypes.

Lury (1996) suggests that issues of imitation and authenticity are important to theorising ethnic urban cultures. Thus, while a symbolic placing of black culture within urban ghetto cultures is a key way in which the white population consumes black culture, authenticity is important for black people who lay claim to an autonomous identity, uncontaminated by white racism. Paul Gilroy argues that music and its rituals provide a basis for such authenticity and, through control over production, circulation and consumption, black styles, music, dress, dance, fashion and languages have had a defining role in shaping popular and consumer culture.

**Figure 5.2 A black music specialist record shop. (Courtesy of Stephen Pope) Source: Lury (1996)**

However, there is a clear 'consuming paradox' at work here. For example, within this dynamic the notion of authenticity acts both as a medium through which white people gain access to the ways of life developed by black people through the stylised use of particular products, and a medium through which black people develop a complex expressive identity. The issue can be considered in a number of interesting ways. For example, Dwyer and Crang (2002) look at the notion of 'multicultural' as a sociological, aesthetic and commercial process. They suggest that commodity culture does not inevitably lead to the production of superficial, thin and bland ethnic differentiations – but that commodity culture can mobilise varied multicultural imaginaries. This suggests that ethnicities are not independent entities that relate to each other post facto, in more or less positive ways, but are the product of a relational matrix in which categories are constructed and applied. Ethnicity is both a complex identity construction and a form of cultural differentiation for the market. However, what is important to understand is that commodity culture can mobilise varied ways of thinking about cultural difference.

For example, one of the most visible ways in which cultural difference and ethnic diversity are represented in contemporary consumer culture is in the proliferation of the availability of foods from around the world. However, Hage (1997) explains that the relationship between urban multiculturalism and the availability of food is not primarily founded on the international migration of people bringing different food cultures to local contexts but rather is based on the growth of cosmopolitan consumption cultures created by international tourism. Hage argues that the proliferation of food cultures represents a multicultural urbanity that is experienced through the consumption of *other* places and peoples. Thus, while business ownership may help migrants secure a feeling of 'being at home', and generate familiarity and a sense of community and belonging in a local context, Hage argues that cosmo-multiculturalism involves the production and consumption of food with a consciousness that these practices are occurring in an international field.

Hage uses the example of Lebanese migrants in Sydney, Australia, who in attempting to carve out a political, economic and cultural life in the city by opening restaurants are subject to international standards of authenticity. Hage argues that savouring culinary experiences is not just a matter of consuming ethnic food but that it is the consumption of difference between ethnic cuisines that is important. In simpler terms, it is by eating the 'world on a plate' that urban dwellers negotiate, manage and consume the ethnic diversity of the city in ways that are prescribed and delineated by the dominant 'hosts'. Hage argues that such local conception and knowledge – of people and food from other parts of the world – is underpinned by a view that those people and places belong somewhere else in the world.

This argument can be explored further in the work of David Parker (2000) who discusses the production and consumption of food and its relationship to the everyday racialised relations in the Chinese takeaways in the UK. Parker shows that takeaways are sites for the enactment of a racial harassment that is bound up with the consumption of difference. He suggests that while the proliferation of Chinese restaurants is widespread throughout the world, a particular British conceptualisation of Chinese food cultures has taken place. Parker suggests that this represents a model of cultural appropriation and transformation which illuminates how cultural difference is bound up with an intimate connection between ethnic food and migrant communities. Parker argues that while foods are often hailed as evidence of openness and multiculturalism, there is a routine racialisation of inequality that is reproduced in everyday transactions. He suggests that this is particularly true of face-to-face encounters. Even where no explicit discriminatory intent is present, it can be seen in the everyday actions and demeanours of customers and workers. Parker shows that Chinese takeaways are governed by a tension-laden objectification of 'race', and takeaway owners and staff make their living on a knife edge that is characterised by novelty and familiarity, risk and comfort. The celebratory multiculturalism often associated with urban life is thus seen to be based on the production and consumption of a negotiated ethnic identity. He thus shows that while the globalisation of Chineseness leads to a complex, shifting, diverse and fragmentary construct that is different in local contexts and for different social groups, what is clear is that this process can generate resilience to the social inequality and cultural stereotyping amongst those who work in takeaways.

In these terms, consumer culture is in part constituted by a range of ethnicities that are reproduced in the production, circulation and consumption of commodities. However, in exploring the entanglement of commerce and culture in the production of ethnicised commodities, Dwyer and Crang (2002) show us that commodification is not something done to pre-existing ethnicities and ethnic subjects, but rather is a process through which ethnicities are reproduced and in which ethnicised subjects actively engage with broader discourses and institutions. Crang and Dwyer look at the 'fashioning' of ethnicities and how designing, making, selling and buying and constructions of 'multicultural' are made up through clothing. They argue that claims of different social groups to inhabit social and symbolic space in the multicultural city are constructed through clothing as commodities. They suggest that we all can (to a greater or lesser degree) interpret and understand how we 'fit' within urban symbolic economies based on clothing. However, Crang and Dwyer argue that while similar clothing styles, designers, labels or types of footwear are ubiquitous across urban and national boundaries there is nonetheless great diversity and nuances in the symbolic economy of fashion in different places.

As such, the diverse mobility of people, objects, images, information and tastes that make up cultural interconnections that reach across the world and which make up people's daily lives in the city are multiple and diverse, and historically and spatially specific. However, what is important is that tracking such transnational flows allows them to be grounded in specific movements of particular people, things and ideas, and this allows their material and symbolic geographies to be examined (Crang *et al.* 2004). Crang *et al.* show that active constructions of identity can in part be constituted by processes of commodification that take place across specific transnational spaces. Such understanding highlights the importance of symbolic imaginaries that circulate through these transnational spaces and how they are materialised not only in terms of different constructions of white appropriations but also in a variety of other ways by diverse ethnic groups in different spaces and places. In simple terms, then, the increasing standardisation of products, generally made by multinational corporations, does mean that 'ghetto' fashion not only is found in poor ghettos in cities across the world, but is worn by young (predominantly white) affluent suburbanites. However, Crang *et al.* argue that there is great diversity, specificity and nuances in the symbolic codes and meanings that different groups ascribe to fashion and clothing. They argue that such consumptive differences ultimately help to tell us a great deal about how consumption mediates and illuminates the differences between different social groups, different cities, and the spaces and places within them.

Let us return once again to the ghetto and consider issues such as white flight and constructions of race in consumer society, particularly in advertising based on racialised and racist representations. Paul Gilroy (1993) argues that consumption is a fundamental mode of opposition or resistance for racial and ethnic minorities on different levels, allowing a reworking of time and space (and night-time style) underpinned by a particular political thrust. Nevertheless, social divisions do not work in such simple terms, and while consumption is an important factor in generating unity and sense of resistance to wider processes of discrimination and inequality, the fact that such identities can be consumed by others is a visual means to show identities can be appropriated.

Zukin shows that in recent years corporate capital has begun to realise that residents of poor districts represent a large market for standard and high-priced brands, and that in the late 1980s and early 1990s certain brands of athletic shoes (Nike) and trekking gear (Timberland) became identified with 'urban' or 'ghetto' cultural styles. For example, an investment partnership with professional basketball player 'Magic' Johnson has brought a Sony Music Theatre into a low-income urban area. Moreover, in the past few years, with reductions in social welfare programmes, local governments and community groups have re-oriented themselves towards attracting mainstream retailers, including supermarkets, in addition to demanding jobs. Although these areas have been purveyors of basic consumer goods, encouraging retail stores fits the general social and political context of reducing government's role and enlarging that of the private sector. The long-term political, economic and cultural effects of bringing new stores and multiplexes into low-income neighborhoods remain to be seen (Zukin 1998a).

It is clear that the growth of urban consumer culture from the early modern city onwards has been profoundly influenced by migration of people from around the world. While it is in cities that the mixing of peoples from around the world has generated hybrid urban cultures that produce both a complex expressive identity for ethnic groups and commodities, those same cultures and commodities can be bought and appropriated by white, and particularly middle-class white, consumers. Moreover, with the growing spatial polarisation of the city along racial and class lines it is those physically, economically and socially marginalised in poor urban enclaves who are excluded and disenfranchised from the dominant bourgeois urban consumer culture. Nevertheless, black clothing styles and music, or Chinese or Indian cuisine, for example, are ubiquitous across urban and national boundaries and there is great diversity and nuances in the symbolic economy of their production and consumption. Despite the structural and everyday inequalities, constraints and prejudices that underpin both consumer culture and urban change, it has been through the dynamics of consumer culture that marginalised ethnic groups have actively contributed to, and have had a profound impact upon, the political, economic, social and cultural life of the city.

## Sexuality

For some theorists, the city *is* the home of the homosexual. It is cities that enable sexualities to be visible. The city is often considered to be 'anonymous', a place of escape by those seeking to transgress conventional boundaries. Cities offer opportunities to those whose lifestyles are labelled deviant or perverse. A growing literature, primarily by and about gay men but latterly including lesbians too, has documented the significance of urban bars and clubs and areas of gay residential gentrification in the establishment of alternative sexual identities during the twentieth century. However, sexuality is not an intrinsically visual phenomenon, and sexual minorities are not necessarily visually recognisable without obvious props of appearance, dress or demeanour. Expressing identity through patterns of conspicuous consumption can thus be of increased significance for the gay community as a visual and potentially political (as well as a personal) statement.

It is in cities that those belonging to the gay male community appear to be relatively unoppressed due to open expression and support of their sexual orientation. One of the key ways to do this is, of course, through consumption. The role of consumption in understanding the relationship between sexuality, masculinity, femininity and consumer society is vital to this assertion. For example, the strength of gay spending is expressed through the 'pink economy' and the fact that gay commerce is an important market. However, certain stereotypes and myths concerning the role of the gay community in urban consumption have arisen. These are most often centred around the idea that gay men have more to spend. This of course belies the reality that there is great diversity in the gay community and not all gay men are white, middle-class and high earners with few financial commitments. Such stereotyping is a distortion of the highly visible parts of the gay community, often those who are young, affluent and professional, and the invisibility of those who are poorer, older and rural. Homophobic abuse and violence are also constant features of urban life (see Gluckman and Reed 1997).

However, despite such shortcomings, much of the work on consumption and sexuality has rightly stressed that social and spatial identities are mutually constituted. This suggests that gay men often do not feel themselves to be gay unless they are spatially visible, and certain areas of particular cities have become well known as gay enclaves. Lawrence Knopp (1995) argues that the density and cultural complexity of cities have led to depictions of sexual diversity and freedom as an urban phenomenon. As a result, minority sexual subcultures and communities often become most closely associated with the city and certain spaces and places within it. However, the concentration of gay cultures in urban space has made it easier to both demonise and control them (and to sanctify majority cultures and spaces). For example, while The Castro in San Francisco, Bondi in Sydney, Canal Street in Manchester, and Brighton are well known as urban gay spaces and archetypes of gentrified gay neighbourhoods, they are also characterised as centres of hedonism and self-indulgence and hence as dangerous underworlds that can be threatening to heteronormative family values.

Jon Binnie (1995) addresses this contradiction in terms of the ways in which the material spaces of the 'pink pound' represent a limited sexual freedom that can often be restrictive. He suggests that queer consumerism and the visibility of spaces of queer consumption in our cities are a powerful assertion of gay economic power; however, they are also a response to a sense of powerlessness. He suggests that shopping offers power and that beautifully designed homes offer a private retreat from the heterosexist (and often homophobic) public sphere, and it is in these terms that consumption is enticing and seductive. However, Binnie guards against overstating the pervasiveness of such an argument. He suggests that for gay men and lesbian women consumption cultures do indeed offer a visible and open expression of sexual identity, but nevertheless these are restricted to particular urban spaces and places. Indeed, openly queer consumption and identity in other places may lead to violence and abuse (see Badgett 2001).

However, what must be stressed is that gay consumption cultures and urban change are not solely contemporary phenomena. For example, George Chauncey (1996) argues that there was an important gay geography in New York in the half-century from 1890 to 1940. Using newspapers, oral histories, court records, books and letters he identifies a public world of gay life in rooming houses, the YMCA and cafeterias, at drag balls, in bath-houses, on the streets and in gay enclaves such as Greenwich Village, Times Square and Harlem in the first half of the twentieth century. He shows that each of these neighbourhoods had different class and ethnic characteristics and public reputations, but all were pervaded by gay consumption cultures. While at this time laws against homosexuality were draconian, they were only enforced irregularly. In these areas of the city, then, networks helped other gay people to find jobs, apartments and relationships. In the city, the sexual adventurer could inhabit a twilight world where sexual encounters could be found in night-clubs, backstreet bars and maze-like apartment buildings.

The explicit relationship between consumption and sexuality is further examined by Frank Mort (1998), who describes the interconnection between sexuality and space in specific urban settings. He described how in the 1980s, Soho in London began to expand through piecemeal and ad hoc development due to the presence of a range of service and media-related industries. He suggests that gentrification was based on Soho's historic legacy of being a centre for the avant-garde – London's own bohemia. He importantly shows that Soho had been a centre for homosexuality since the 1920s, and that female prostitution had an even longer presence in the area.

Mort shows that the pubs and clubs adjacent to Leicester Square provided one important anchor point for these sexual cultures. He argues that by the 1950s and 1960s the youth fashion scene generated sexually ambiguous clothing that attracted a wide range of heterogeneous bohemian and ethnically diverse people. Though they were differentiated in terms of their public visability, economic leverage and levels of social acceptance, there were different cultures linked by consumption. Mort shows how today this legacy has had a particular impact on Soho. One of its main thoroughfares, Old Compton Street, is promoted as the capital's gay village. In this area the consumption services to gay men are extremely diverse, from restaurants, cafés, boutiques and hair salons to sex shops, and what is promoted is a vision of an exclusively homosexual life and lifestyle fulfilled through consumption and serviced by homosexual entrepreneurs.

It is in such terms that contemporary gay consumerism is often characterised as being constructed around hedonistic lifestyles (Hennessy 2000). This is, however, at the exclusion of other ways of acting and being. Mort suggests that constructions of gay hedonism represent a powerful consumption–space identity couplet. However, Binnie (2004) is more critical about the ways in which consumption in urban areas is constructed around particular identities, activities and ways of behaving. He argues that the development of 'gay villages' undermines conflicts that surround the confusion and challenge of urban space. He suggests that the citizenship being promoted as part of this process is a disciplining and normalising agenda that ultimately reinforces exclusionary discourses that surround boundaries in cities. Binnie argues that quarters fix queers in place – ascribing identities and consumption practices that are constraining and ultimately produce a less dynamic lesbian and gay culture. He argues that in gay quarters difference is annihilated (see Case study 5.4).

## Case study 5.4 **Cosmopolitanism, consumption and gay space**

The contemporary city is often argued to hold new opportunities for previously marginalised groups, while simultaneously dividing them. Binnie and Skeggs (2004) show, for example, that as new markets for leisure consumption with new forms of branding and the territorialisation of space have developed, lesbian and gay urban space has increasingly been marketed as cosmopolitan spectacle. However, they argue that underpinning this process are certain divisions, and they seek to unpack such conflicts by considering who can use, consume and be consumed in gay space.

Binnie and Skeggs show that the sexual politics bound up in the development of spaces such as urban gay villages are focused around notions of cosmopolitanism

continued

– commonly conceived or represented as a particular attitude towards difference. To be cosmopolitan one has to have access to a particular form of knowledge, and be able to generate knowing from this authority. In these terms, lesbian and gay space is cast as cosmopolitan, and is based around the idea of a 'global gay' identity that does not adhere to national boundaries and that generates global consumers who contribute to transnational flows of capital and culture. The Mardi Gras in Sydney, European Pride in Manchester, international travel guides, pink papers and magazines, and gay urban space formed around design-led bars, restaurants, boutiques and brasseries are the 'stuff' of this globalisation. Moreover, such events, locations and landmarks have also been commodified in official urban place promotion and tourism campaigns.

Binnie and Skeggs argue, however, that central to this commodification is an essentialisation of 'gayness' based upon middle-class bourgeois consumption cultures. This, of course, excludes lesbian and gay people (both in representation and in the real city) who are working-class, older, from ethnic groups, and who constitute gay sexual subcultures – in sum, those who do not necessarily live up to the image of young, glamorous and trendy urbanites. Gay cosmopolitanism is thus used to promote gay neighbourhoods as a non-threatening authentic commodity in order to broaden its appeal to consumers regardless of sexuality, hence attracting customers and money. If you are sophisticated and cosmopolitan you can legitimately occupy these spaces, and middle-class gay men and lesbians can safely inhabit the space in which for a short time they become a politically neutral motif of regeneration.

Source: Binnie and Skeggs (2004)

Binnie (2004) shows that as consumption is now central to how citizenship is defined, the management and disciplining of the self occurs through choices we make as consumers. The growth of visibility of lesbian and gay men associated with the 'gay marketing moment' and development of the pink economy discourse has facilitated the articulation of rights claims, but has also spawned debate about the nature of the freedoms won and the exclusions produced. Inclusions and exclusions are based around ability to consume. Gay commercial spaces are bounded and commercial whereas queer counter public (for example, spaces for public sex such as toilets and cruising areas) are potentially more democratic and are not subject to commodification.

One final issue that needs to be raised is the ways in which different cities accommodate, tolerate or celebrate lesbian and gay people. For example, Sally Munt (1995) vividly describes her different experiences of the UK cities of

Brighton and Nottingham in terms of her lesbian identity. In contrast to a more homo-friendly street and institutional culture in Brighton, Munt lamented a restricted sexual citizenship in Nottingham where she experienced a relatively more oppressive and homophobic urban culture. Similarly, Binnie (2000) discusses how in Bolton, a predominantly working-class town in the north-west of England, seven men were arrested and subsequently prosecuted for consensual same-sex activities in their homes in 1997. Such examples identify that Bell and Binnie (2000) are correct to assert that the city is a prime site both for the materialisation of sexual identity, community and politics and for conflicts and struggles around these. In sum, cities (and spaces and places within them) are different from one another, and the treatment of sexual diversity and sexual citizenship is constructed differently in different cities.

## Subcultures

Miles (2001) argues that if there is one social group that could generally be considered to be full members of consumer society, then it is young people. Young people emerged as an important consumer entity in the 1950s, when 'adolescence' as a distinct period of our lives was formulated. Chatterton and Hollands (2002) show that while the Second World War was an important turning point in consumer society in general, with post-war economic growth bringing employment and higher incomes, it was particularly important in terms of expectations amongst young people. Youngsters became 'affluent teenagers' and had more money to spend. Hence the teenage market developed in the 1950s and 1960s around fashion, music, food, clothes, drink and drugs, and since that time teenagers have been at the forefront of conspicuous, fast-paced, leisure-oriented consumption. From the teenager onwards, punks, teddy boys, mods, disco, acid house, rave, gangstas and goths are just a few classic examples of youth subcultures. Such subcultures each try to rebelliously establish their identities in opposition to the clothing, activities, values (and so on) that they conceive of as staid, 'square', 'establishment' or simply those that characterise the older generation.

One of the first writers to discuss the sociological importance of youth culture was Stanley Cohen, who described the 'moral panic' that surrounded fights between mods and rockers in the UK. Following this work was the development of the concept of youth subculture; Dick Hebdige's book *Subculture: The Meaning of Style* (1979) described youth consumption as a political phenomenon and how 'style' was imbued with significance. For example, Hebdige argued that working-class teddy boys reacted against dominant culture and against a society structurally subordinating them in relation to economic position, unemployment, housing, and so on. He suggests that youth cultures often adopt a process such as 'bricolage' whereby mainstream symbols are re-assembled to form differing and often counter-cultural meanings perceived as political by marginalised working-class teenagers. Nevertheless, despite this political content subcultural style is more often than not incorporated into mainstream culture (for example on catwalks) or is commod-ified, packaged and sold back to young people.

A major and more recent impact on the understanding of the processes that underpin the sociology of subcultures has been postmodern theory. Culture and consumption, it is argued, are central to the argument that youth subcultures have become neo-tribes (Maffesoli 1992). This suggests that there has been an erosion and loosening of traditional sources of identity and a blurring of traditional social relations such as class, gender, ethnicity, sexuality and national identities. This suggests that it is difficult to tie youth cultural styles down to any strict notion of social class. It follows, then, that there is little revolutionary opportunity offered by subcultural style delineated not by social class but by aesthetic belonging (Maffesoli 1992).

However, Chatterton and Hollands (2003) suggest that contrary to either a free-floating, individualised 'pick-and-mix' story of postmodern youth cultures or a simple 'class correspondence' model of leisure, contemporary youth consumption is instead characterised by hierarchically segmented consumption groups and spaces in cities. They use the example of night-time youth consumption, which they suggest is highly structured around drinking circuits or areas, each with its own set of codes, dress styles, languages and tastes. Despite postmodern contentions that hybrid eclectic styles today make it more problematic for young people to distinguish between themselves and other youth cultures, many young people have no such difficulty in identifying varied nightlife consumption spaces – mainstream, residual and alternative – that are inhabited, made and remade by a range of competing youth cultural groups characterised by various cultural tastes and traditions.

For example, Chatterton and Hollands (2002) show that youth culture has devel-oped an important presence in urban cityscape around certain niches – for example, mainstream taste communities around gentrification and cosmopolitanism. These include café culture and 'style venues' with polished floors, minimalist or branded décor and designer drinks. However, the working class, the unemployed and welfare-dependent and criminalised youth cultures represent the 'other city' of dirt, poverty, dereliction and violence in contrast to the gentrified mainstream. Residual youth groupings, including young unemployed, homeless, poor and often those from ethnic backgrounds, are excluded, segregated, policed and in some cases swept off the streets. Hence, younger generations of the socially excluded have limited choices of nightlife: ghettos, community pubs and centres, social clubs, house parties and the street. Groups such as these will continue to be maligned and increasingly excluded from city-centre nightlife, as gentrification and urban rebranding for the wealthy elite continue.

There are also alternative nightlife spaces, usually independently owned, single-site music clubs and bars. Chatterton and Hollands also point to the importance of house and free parties in providing for the particular needs of youth identity groups based around styles of music (and clothes), such as garage, nu-metal, hip-hop, indie, grunge, and so on. They further argue that such marginal spaces distinguish themselves from corporate venues by the presence of people with body piercings and tattoos as well as highly stylised, sexualised or physically dangerous dancing.

However, minority cultural identities and neo-tribes and subcultures do not negate the idea that social and spatial divisions, inequalities and hierarchies continue to exist in our cities. At night, internal cultural diversification in the night-time economy is as much about social hierarchies and securing continued profitability as it is about hybridity. Urban youth night-time consumption cultures remain segmented around the commercial mainstream, with its various subdivisions and diminishing opportunities for alternative and residual experience. However, what is clear to Chatterton and Hollands (2002) is that such consumption groups and their spaces are fluid and overlapping rather than rigid. They argue nonetheless that there are still profound structural influences on people's lives and that youth identities and lifestyles are as much underpinned by social divisions and transitions as any others.

Steven Miles (2001) further investigates the relationship between young people and urban consumer society. He argues that observers of the consuming city might come to the conclusion that, more than any other age group, young people are happy to indulge in the attractions of consumer lifestyles; however, they are at once both upholders of the consumerist ideal and usurpers of urban space. He suggests that the young people are often portrayed in a negative light – as rebellious, deviant and drug and alcohol abusers, and as often violent defenders of their 'patches'. However, Miles argues that to view the construction of local youth identity solely in terms of a reaction to young people's perception and experience of being marginalised in the city would be misleading.

Miles suggests a more critical approach and argues that young people's use of urban space may serve to reinforce the status quo. He uses the example of how young people negotiate CCTV surveillance of their activity. He shows that young people move through concealed urban spaces and invisible routeways that allow them to re-appropriate urban space. He shows that their resistance is limited, and while young people are highly visible in urban areas we need to avoid romanticising young people's consumption practices. Miles argues that young people's use of urban space represents an effort to secure space within dominant power structures rather than outwit them.

As such, while young people express themselves through their lifestyles, which are simultaneously individualised and communal (consumption is an individual and subcultural safety net), this ultimately reinforces the ideological underpinning of the market. Miles argues that teenagers take it for granted that they were 'born to shop', and that we must not exaggerate the rebelliousness of young people. He suggests that most young people are happy to take up an active and knowing role in contemporary consumer society. Consumption, or at least the idea of consumption, to young people is a means to express their sense of difference while simultaneously allowing them to feel that they belong. While gender, ethnicity, class and age all play an important role in urban experience – young people's experience of urban life is diverse – whether young people 'hang out' in shopping malls or on the street corner, this only represents surface resistance.

Miles concludes that young people have very few places to go in our cities. He suggests that where young people do 'hang out' it is an attempt to re-appropriate public space in order to create meaning and context in their everyday lives. However, Miles suggests that this (temporary) occupation of parts of the city is less to do with resistance and more to do with making space for themselves. This engagement is thus bound up in the consumer society that allows young people to develop a sense of freedom through which they can construct their own meaning, but which ultimately fulfils dominant social order (see Case study 5.5). However, young people are not necessarily unhappy about their situation and hence they not only use urban environments in pro-active ways but are well equipped to deal with urban consumer culture.

## Case study 5.5 **Skateboarders and the city**

Skateboarding has over the past two decades become one of the most high-profile and visible ways in which young people engage with urban life. With its own semi-conventionalised rules of behaviour, language, clothes, magazines, videos and music, skateboarding creates a comfortable sense of security in the face of the threatening insecurity of the modern world and city. While skateboarding is frequently non-organised or semi-organised, it is generally a male-dominated activity. However, to understand skateboarding fully it must be seen in relation to its spatial constituents. Skateboarding is not just socio-cultural but is physical, undertaken against the material of the modern city. It is in this way that the urban practice of skateboarding implicitly and continually critiques contemporary cities.

**Figure 5.3 *Skateboarding in the city. (Courtesy of Iain Borden)***
***Source: Borden (2001)***

**Figure 5.4 *Benches designed to deter skateboarding.***
***(Courtesy of Iain Borden)***
***Source: Borden (2001)***

Through its everyday practice skateboarding suggests pleasure rather than work, and the city is the hardware (skate parks are not everywhere and are often poor). The rich architectural and social fabric of the city offers skateboarders a plethora of buildings, social relations, and time and spaces for skateboarding. Thus, while commercialisation pervades every aspect of urban life, skateboarders pursue a style of experience not beyond formal styles of architecture and commodified lifestyles of fashion and food, but adapt and manipulate the 'proper' use of the city, often in highly symbolic sites.

However, what is clear is that while skateboarding does not offer an explicit political critique, nor does it promote social disruption, it is a diffused and dispersed activity which often conflicts with both public and private institutional interests at local levels. For example, there have been increasing anti-skate measures, such as police 'zero tolerance' and skateboard confiscations and CCTV surveillance. Moreover,

continued

design features that hamper skateboarding such as curved bus benches, window ledge spikes, rough textures and surfaces, spikes and bumps added to handrails, blocks of concrete placed at the foot of banks, and chains across ditches are increasingly common features aimed at deterring skateboarding. Nevertheless, the city thus becomes an instrument for skateboarders who see it as a series of micro-spaces rather than as a comprehensive urban plan, monument or grand project.

Source: Iain Borden (2001)

## Conclusion

This chapter has shown that consumer culture is not solely concerned with the purchase and display of consumer goods but is part of the making of culture itself, the legitimation practices it produces, and its aesthetic content in specific socio-spatial settings. As such, consumption can be considered as one of the ways in which *social structure is mediated to and by individuals*. However, while this explains how consumption is central to urban change, its manifestation in specific socio-spatial settings needs more investigation (Wynne and O'Connor 1998). Thus, studies of consumption play an important role in our understanding of how identity and selfhood are constructed, highlighting how the relationship between power and resistance is played out through consumption. However, what becomes clear is that while these issues can be theorised at a general level, they can only be more fully understood and established with reference to empirical groups and contexts. As such, there must be a conscious attempt to ground accounts of emergent lifestyle groups in terms of both place and space.

**Learning outcomes**

- To have an understanding of the relationship between consumption and identity
- To be able to describe the relationship between consumption and constructions of class, gender, ethnicity, sexuality and subcultures
- To offer insights into how the relationships between identity, power and resistance are played out through everyday consumption
- To be able to critically appraise how class, gender, ethnicity, sexuality and subcultural style are constructed differently at different times and in different places

## Further reading

David Bell and Gill Valentine (eds) (1995) *Mapping Desire: Geographies of Sexualities*, London: Routledge. A landmark book that investigates the relationship between the city and sexuality, from the local to the global context, around sexual identities and through urban sites and rituals of resistance.

Zygmunt Bauman (1998) *Work, Consumerism and the New Poor*, Buckingham: Open University Press. An important book that investigates the often ignored issue of contemporary urban poverty through topics and issues such as the welfare state, the underclass, life and work ethics and consumption.

Paul Chatterton and Robert Hollands (2003) *Urban Nightscapes: Youth Culture, Pleasure Spaces and Corporate Power*, London: Routledge. An interesting and accessible book that has highlighted the complex world of the city at night and investigates topics such as students, regulation, corporatisation, social exclusion and 'alternative' cultures.

Les Back (1996) *New Ethnicities and Urban Culture: Racisms and Multiculture in Young Lives*, London: University College London Press. An influential book that provides great insights into contemporary constructions of urban ethnic experience. Racism, community, music consumption, resistance and neighbourhood nationalism are just some of the substantive issues covered in this book.

# 6 Consuming the city

**Learning objectives**

- To look at the relationship between the physical city, representations of the city and everyday experiences of urban life
- To show how official representations of the city seek to construct a particular image of physical, political, economic, social and cultural competitiveness
- To think about the diverse and varied representations of the city in popular culture
- To describe the relationship between the imagined and the 'real' city

This chapter shows that the city, and spaces and places within it, not only are sites of consumption but are also themselves consumed. Representations of the city in the virtual world of consumer society – in films, books, magazines, advertising, fashion and songs – and the ways in which the city is represented in place promotion, planning and other official discourses are discussed. The chapter outlines the ways we consume spaces and places (visually and through smell, sound and touch) and how we interpret and experience the city. These topics are framed through the work of Henri Lefebvre (1971) and Edward Soja (1996), who highlight how the imaginary city and the ways we consume urban spaces and places (in relation to a complex matrix of identity positions noted in the previous chapter) not only affect the way we experience the city, but also inform the material development of our cities.

Subsequent sections will investigate, through specific examples, Lefebvre's and Soja's argument that the production and consumption of the city are bound up in the interplay between the physical, imagined, perceived and experienced city. There are, of course, an enormous number of different ways that the city is represented in popular and institutional forms, and this chapter seeks to provide diverse and varied exemplars – to unpack in detail how the city is itself consumed in a variety of ways through a number of specific examples. First, the way in which we differentially consume urban spaces (such as the street) will be discussed. Secondly, the production and consumption of place-promotion campaigns will be investigated. Finally, the impact of popular culture forms on the city, with a focus on the movies, will be unpacked. The chapter begins, however, with a discussion of how the city can be considered as a 'text' that can be read and interpreted, followed by an introduction to the work of Lefebvre and Soja, who show that the city is both produced and consumed through the interplay of three spatial realms.

## The production and consumption of the city

Urban studies has moved away from a sole focus on the city as a subject to be measured through scientifically hypothesised approaches towards a more qualitative understanding of urban political, economic, social, spatial and cultural practices and processes. One of the most important approaches to emerge from this theoretical understanding is the idea that the city is a 'text' that can be 'read'.

For example, John Rennie Short (1996) shows that cities are produced and repro-duced in a number of different ways, including the interface of (and resistances to) capital and power. Of similar importance is the production and consumption of the symbolic representation of the city in myths, ideologies and images. Short argues that the city can be seen as a set of signs – a non-verbal system of communication. The city is thus a container of messages that we are all exposed to, and the 'writing' of the city involves the 'reading' of the city. Nevertheless, there is no one-to-one correspondence between the production of the message and its consumption – multiple readings vary across society.

This argument is underpinned by the idea that the social world is made up of 'signs' that can be read via the discipline of semiotics – first developed by Swiss linguist Ferdinand de Saussure (1915). It was not until much later, however, that the potential of semiotics was more fully developed through the work of Roland Barthes (1965, 1982). Barthes showed that the reading of signs through semiotic interpretation could be achieved through an analysis of denotation and connotation. Barthes described denotation as the 'uncoded' meaning of the sign, and gave the example of a bicycle, as a means of transport. He also argued that at the level of connotation the sign acquires social and cultural meaning according to how it is depicted. Hence, a bicycle could have many connotations or meanings: as a sign of health and fitness or associated with the countryside in opposition to the car or town. However, Barthes highlighted a more particular cultural association, and showed that in its relationship to the Tour de France, the bicycle has come to symbolise Frenchness and notions of French national identity. Barthes called this kind of connotation a 'myth', which he saw as akin to a cultural stereotype (see Brooker 1999 for a further description).

More recently, Stuart Hall (1973) took up the method of semiotics as a way of interpreting visual communication, arguing that all images are both encoded and decoded. Hall suggested that all images are encoded in the production process and also via their placement within a certain cultural setting. He also highlighted how images are consumed through a process of decoding, whereby viewers and readers interpret those images. Hall described three ways in which images are interpreted; first, through a 'dominant-hegemonic' reading, where people identify with the dominant message that underpins the image or text in an unquestioning manner. Second, is the negotiated reading, where people accept but question the dominant meaning of the image. And finally, Hall described an oppositional reading, where the reader or viewer of the text completely disagrees with the ideological position embodied in the image, or rejects it altogether, or simply ignores the message.

Hall's depiction of three ways of interpreting the messages and images has to a large degree been superseded by the view that there are multiple ways of consuming images and texts. However, his contribution to the realm of semiotic interpretation, initially proposed by Saussure and later developed by Barthes, has had an important impact on urban theory. Such advances generated an interest in deconstructing the meanings of the multiple signs and images that constitute the city, and led to work that sought to describe the city as a text to be 'read'. For example, Gottdiener and Lagopoulos's (1986) book *The City and the Sign* provides an excellent example of the interpretation of the social production of the city through semiotics. The book embarks on a project to 'read' and interpret the city, including study of urban architectural styles; street names and signs; shop signs and the commodities sold within; monuments, memorials and statues; motorways; place-promotion campaigns; street fashion, ways of life, gender roles and subcultural style; and, of course, shopping malls.

The rationale underpinning urban semiotics is to make power relations visible. Gottdiener and Lagopoulos thus argue that while power is written large across the urban landscape it remains invisible, and moreover, they seek to depict how people interpret, accept or resist such power. There exist multiple, contested changing readings and relations of power. King (1996) thus argues that urbanism should not be taken for granted, and that straightforward social and economic and political analysis is inadequate. To look at the culture of the contemporary city is not therefore simply about the subject of representations constituted through different categories of knowledge – and our experiences mediated by gender, ethnicity, class, sexuality, nationality, ideology, and so on – but about how different subjectivities affect not only experience but physical, economic, social and cultural change in the city. Thus, while the city read as 'text' is a useful way of interpreting and analysing representations of urban life, theorists have sought a more satisfactory way of explaining how representations of cities, the physical city and everyday experiences of the city are intertwined.

The most coherent theoretical approach to addressing this difficult goal has been through the work of Henri Lefebvre (1971) and, later, Edward Soja (1996). Soja depicts space as being discursively constructed as real, imagined and perceived – a 'trialectics of spatiality'. Soja's work borrows extensively from Lefebvre, and a plethora of other theorists, in an overture to thinking about the meanings and significance of space. Soja seeks to explain the ways an inherent spatiality of human life is constructed – where meanings such as location, locality, landscape, environment, home, city and region are conjured from.

Soja uses three main concepts to promote an understanding of the social production of spatiality. First, 'spatial practice' – defined as a spatiality which 'embraces production and re-production and the particular locations (milieu specific) and spatial sets (ensembles) characteristic of a spatial formation'. It ensures continuity and some degree of cohesion and 'a guaranteed level of competence and a specific level of performance' (Soja 1996: 10). Spatial practice is the process of producing the material form of social spatiality – it is thus the medium and outcome of human activity, behaviour and experience. This is the materialised, socially produced, *empirical* space of houses, streets, city squares, parks, roads, shops, car parks, cities, regions and countries. This is 'firstspace', the spatial practice under capitalism, which links the repetitive routines of everyday life, and the routines, networks, workplaces, private life and leisure of its inhabitants. Secondly, in Soja's terms 'representations of space', is 'a conceptualised space, the space of science, planners, urbanists, technocrats, artists' (Soja 1996: 11) – all of whom identify what is lived and what is perceived and conceived. This is 'secondspace', and it is often utopian in thought and a vision that is semiotic or encoded. This is spatiality represented in official documents, plans, books, paintings, songs and academic writings. These representations are important as they inform the way we think about spaces/places, and thus can have material consequences for the fortunes of those spaces/places. Thirdly, Soja depicts 'spaces of representation' as a 'complex symbiosis' of both the above examples. This 'thirdspace' is the space that is directly lived and experienced. This is particularly important in understanding relationships between power and space – and relations of dominance, subordination and resistance. 'Thirdspace' is the ways in which we experience space and the ways we feel 'at home', happy, threatened or in danger in particular spaces/places.

Soja's conception owes much to Lefebvre's (1971) work, which also considered that the production of space is manifest in three realms that he described as a 'spatial triad'. First is 'representations of space', which Lefebvre considers to be the physical city – streets, houses, roads, malls, urban villages, monuments, parks, factories, skyscrapers, and so on. Lefebvre shows this space to be constructed by professionals and technocrats, such as planners, developers, engineers, architects, urbanists and geographers. This space is made up of signs, codifications and objectified representations, and Lefebvre suggests that it is a space conceived through ideologies, power and knowledge. He suggests that power is embedded in this representation of space in order to impose signs, codes and order.

Secondly, Lefebvre describes 'representational space', as the directly lived space of everyday existence. He suggests that this space is made up of the complex symbols and images produced by its inhabitants and users, and that representational space overlaps with physical space (representations of space), and by making symbolic use of its objects (streets, houses, roads, malls, monuments, factories, and so on) people make sense of their everyday lives. Representational space is felt rather than thought, and Lefebvre argues that it is an elusive space – although it is experienced by people in their everyday lives – and while ordered space tries to intervene in and rationalise the way that people experience and perceive the city, it eludes such rationalisation. Finally, Lefebvre describes 'spatial practice' as the ways in which people structure their everyday lives within broader social and urban realities. These include routes through the city, for example, or networks and patterns of interaction which link places set aside for work, leisure and play.

The work of Lefebvre and Soja provides a conceptually important and convincing way of understanding that the city is at once real, imagined and perceived, and that it is through a combination of these elements that the city exists. Merrifield (2000), however, argues that Lefebvre is vague about the precise manner in which spatial practices mediate between the conceived and the lived. Such shortcomings perhaps help explain Soja's reworking of the three realms that constitute the social production of space in order to more fully elaborate on the relationship between the physical city, the city represented in signs and the experienced city.

Nevertheless, when taken together, urban semiotics or the more holistic concepts of the social production of space as developed by Lefebvre and Soja provide a coherent theoretical framework that can be utilised to unpack the construction and experience of urban life. However, despite this theoretical advance, research has generally tended to overlook the practices and values of consumers and privileged the analyst's own reading of the signs and symbols of contemporary urban consumption (Jackson and Thrift 1995). This failure to engage with the multiplicity of consumption cultures (and so much of their power and dynamism) is only now beginning to be rectified. Similarly, those seeking to link production with consumption, to identify consumptive subversions and the different meanings which different people assign to particular activities and practices (and how local and national states mediate this), are beginning to show how such issues are historically and geographically constructed and negotiated. In essence, research has tended to ignore Lefebvre's 'representational space' or Soja's 'thirdspace' – the ways in which space is directly lived and experienced, and the discourses and structures of power, domination, subordination and resistance which mediate everyday life and come together to create the physical city, and its 'real' and 'imagined' worlds.

The remainder of this chapter will address these theoretical concerns and look at the ways in which the city is itself consumed. Sections will investigate the consumption of the physical city (spaces of representation), consumption of the imagined city (representations of space) and how people consume cities (experienced space). The discussion will reflect on how these spatial dimensions are intimately bound together, and together inform the development of our cities.

## Consuming the real and imagined city

Chapter 4 showed how people engage with urban landscapes in 'ordinary' and mundane ways (de Certeau 1984; Lefebvre 1984). However, urban semiotics, and the conceptualisation of the social production of space, highlight that while the city is a shared text there is no equality in reading or writing the city. As Chapter 5 showed, between the production of urban form and its consumption (in the realm of everyday life) fall the intervention of multiple and contested readings and the mediating factor of different social groups' relations to power. This section seeks to unpack this relationship by elaborating on the trialectics of spatiality through specific examples. The focus in this section will be on the production and consumption of constructions of the physical city, and how these overlap with the experienced, imagined and perceived city.

However, in order to highlight how the production and consumption of the city are mediated through different spatial realms there are many examples and case studies to draw upon. Studies of urban architectural styles, street names, statues, motorways, street fashion or the shopping mall all offer a wealth of material to embark on this project. However, one of the most effective ways to elaborate the argument is to make the familiar strange, through the study of people looking at ordinary things in extraordinary ways. This highlights in an 'extreme' way how we all consume the city in our everyday lives.

Such a conception of the relationship between consumption and everyday life is undertaken in the work of John Urry (1995), who is interested in the social relations of place – in essence, how people experience spaces and places. Urry is particularly interested in how social relations and interactions are related to consumption, specifically in terms of the interdependency between the consumption of material objects and the natural and built environment. Urry (1995) argues that the relationship between the social relations of place and consumption is important in four ways. First, places are increasingly sold as centres for consumption, and hence provide the context within which goods and services are compared, evaluated, purchased and used. Secondly, places themselves are in a sense consumed, particularly visually. Thirdly, places can be literally consumed; what people take to be significant about a place (industry, history, buildings, literature and environment) is over time depleted, devoured or exhausted by use. And finally, it is possible for localities to consume your identity, so that places become literally all-consuming. This can be true for visitors or locals, and works in a way that places can produce multiple enthusiasms.

For instance, local people might utilise urban public space, such as squares or plazas, in a wide range of everyday ways that are divorced from their historic or political context – such as for playing sports, picnics, and for more transgressive activities such as public sex or drinking. Tourists, on the other hand, might well judge such activities as 'out of place' within spaces that for them are experienced through a grounding in an understanding of the historic, political and aesthetic worth. Despite such conflict, this urban public space provides consumption activities for both locals and tourists, and shared if multiple pleasures. Such contentions highlight the complex ways in which place and the consumption of goods and services are interdependent.

Urry (1995) stresses how images of place are routinely used in the symbolic location of goods and services (in, for example, tourist advertising campaigns and civic place-promotion campaigns – this will be returned to later). Similarly, living in or visiting particular places often entails certain kinds of consumption (such as eating doughnuts and playing slot machines at the seaside, buying pottery in Stoke-on-Trent or fashion in Milan, eating hot dogs in New York, and paella in Spain, and so on). It is through such examples that Urry shows how certain kinds of products and services can only be obtained by visiting a particular place. Finally, Urry argues that images of places are constructed out of particular products and services which are available in particular places.

These examples of product–place associations ultimately serve to show the complex interdependencies between consuming goods and services and places (Molotch 2003). Moreover, what links them together is the patterns of social life organised in and through particular places, ways of walking, talking, social relations associated with the seaside or old industrial cities. Urry argues that while such patterns are significantly commodified, there is generally a complex mixing of commodification, the selling of those atmospheres and the remaining associations of local people. Urry elaborates his argument most fully in terms of tourism. He suggests that to look at and gaze on particular objects such as piers, towers, old buildings, artistic objects, food, countryside, and so on is one of the obvious things that tourists do. Other consumption activities, including the purchase and use of hotel beds, theatre tickets, deckchairs and loungers, are also central to the tourist consumption act.

However, Urry (1995) argues that the tourist gaze remains a potent consumptive act. And he argues that to look individually or collectively upon aspects of landscape or townscape that are distinctive offers an experience that contrasts with everyday life. It is through the tourist gaze that places are interpreted through fantasy, intense pleasure and often on a different scale from everyday experiences. Such consumption is fuelled through anticipation and is constructed through a variety of non-tourist practices such as film, newspapers, TV, magazines, records, video and word of mouth that all help to construct the gaze. However, the city is not only produced and consumed through our collection of signs, depicting landmarks such as the Eiffel Tower in Paris, New York's skyscrapers, Las Vegas's casinos, India's Taj Mahal and the English pub. The city is produced and consumed through commonplace and everyday spaces and activities in all cities throughout the world, in parks, the street, shopping centres, and so on.

## The street

One way to elaborate on this complex relationship between consumption and place is to look at a specific example of a significant tourist site that is also a site of everyday consumption – making the familiar strange. One good example is highlighted by Peter Jackson (1998), who suggests that since the nineteenth century, if not before, 'the street' has been regarded as a lively and contested public domain, the site of popular protest and political struggle. Similarly, Valentine (1998) discusses the relationship between consumption and streets. She suggests that in western cities there has been a long 'civilizing process' in defining what it is acceptable to do in the street and when behaviour is out of place. For example, in medieval times it was commonplace for people to eat, belch, fart, spit, shit, and so on in public. Today, such antisocial behaviour would of course be frowned upon.

However, as interesting is the way in which not only are the production and consumption of food now highly regulated but there are also social conventions about eating in public in many western cities. The western street is characterised by relative order and segregated residential and commercial or retail activity.

As Valentine (1998) shows, eating on the street can inspire shame and embarrassment in ourselves. Public eating is also seen as inappropriate because it has the capacity to revolt or embarrass those who witness this intimate activity. Good middle-class taste does not include eating on the street; it is considered as 'common' or 'uncouth'. However, Valentine (1998) also argues that given the wide range of hybrid food available to eat on western streets there is now a wide range of public food cultures jostling for space (high/low/ethnic, etc.) and this has contributed to a relaxation of some of the social codes around performances of the self. Such indication of diverse consumption cultures is also reflected, for example, in the very different ways in which codes of dress and ways of acting and eating have changed over time and are seen to apply differently to tourists and non-tourists. This is highlighted through the different types of clothing and ways, routes and speeds of walking between tourists and non-tourists. Other examples include tourists eating in, or taking photos of, what for locals might seem unusual places to eat, such as graveyards or at the side of the road, or seemingly ordinary things and buildings to photograph. Such examples serve to highlight Urry's argument about the specific consumption activities involved through the tourist gaze, and more everyday uses of space.

Despite such diverse and multiple consumption cultures that undoubtedly exist in western streets, there has nevertheless been a standardisation of the production of consumption. Shops, for example, tend to be uniformly organised in neat blocks and buildings. The street itself can be either pedestrianised or given over to traffic, but streets remain highly regulated spaces where legal or social conventions delineate good behaviour. Moreover, urban planning policy ensures particular usage of urban space (described earlier in this book). High-street shopping, for example, tends to be characterised by chain stores, banks, building societies, food outlets, bars and restaurants as well as a few unregulated street traders. However, it is clear that while there has been a domestication, standardisation and civilisation of western streets, the ways in which people consume such spaces is rich and diverse (see Case study 6.1, p. 163).

In contrast, Edensor (1998) looks at consumption on Indian city streets in terms of the sensual and social experiences of space. He argues that cities cannot simply be read as texts, but must be passed through, to fully understand the experience of the people living in, working on and visiting streets, to interpret their experience through social, sensual and symbolic processes. Edensor shows that Indian streets

**Figure 6.1** *The 'civilised' street. (Photo: Mark Jayne)*

are labyrinths, with numerous openings and passages. They are not just for shopping but the site for numerous activities which co-exist alongside each other. These include workplaces, schools, eating places, transport terminals, bathing points, political headquarters, offices, administration centres, places of worship, and temporary and permanent dwellings. He describes a mixture of overlapping spaces that merge public and private space, work and leisure, and holy and profane activities. He shows a mixing of dentists, fortune tellers, shoeshiners, barbers, letter-writers, shoe repairers, bicycle fixers and tea-wallahs, as well as mobile stalls of all kinds, engineers, smiths, potters, bookbinders, metal workers and other industrial spill-outs on to the street, all of which further blur frontstage and back-stage, where the production and marketing and selling of goods take place in the same place, and often by same person.

Edensor describes how the street is a social space for gossip, a place for drinking tea, a place where adverts are visually paraded or broadcast through loudspeakers. Political and religious processions and demonstrations are commonplace, as are entertainment, musicians, puppeteers, magicians, hawkers, beggars and holy men occasionally performing acts of abstinence and endurance. In the Indian street there is a constant stream of temporary pleasurable activities, entertainments and transactions as well as more mundane personal activity such as loitering with friends, sitting and observing, and meeting people.

**Figure 6.2  Cooking on an Indian street. (Courtesy of Tim Edensor)**
*Source: Edensor (1998)*

**Figure 6.3  A barber's on an Indian street. (Courtesy of Tim Edensor)**
*Source: Edensor (1998)*

Edensor highlights that movement in Indian cities is important. He suggests that you cannot walk in a seamless uninterrupted journey but only through a sequence of interruptions, and that rapid progress is usually frustrated. He particularly focuses on the rich sensual encounter that is part of moving through the Indian street. Edensor describes motion against a backdrop of randomly arranged buildings and objects that provide surprising and unique scenes. Such haphazard features and events dis-order the gaze and spatial regularity. Different textures brush against your body, and the smells of the Indian street are rich and varied, with a jumbled mix of pungent smells – sweet, sour, acrid and savoury. Equally diverse is the soundscape, which combines the noises generated by numerous human activities, animals and forms of transport with performed and recorded music, to produce a symphony of diverse pitches, volumes and tones. The Indian street thus offers an environment of rich and diverse smells, sounds and indefinable 'atmospheres'. Late capitalism has rendered street life predictable, marked by sensual deprivation, with difference reduced to commodified sameness. The destruction of the functional and cultural diversity of the street has thwarted human contact, the desire for difference and the need to wallow in the obscure and confusing. This is contrasted to the few disorganised tourist spaces and examples of undomesticated street life in western cities, such as fairs and carnivals which at best can be described as organised dis-order.

It is clear that the examples of a western street in comparison with an Indian street help to highlight the different ways that they are produced and consumed. Each provides an understanding of the ways in which the street is differentially physically structured, is brought to life through very different social relations and forms of sociability, and hence experienced differently by people with very different social backgrounds. This contrast helps us to understand Lefebvre and Soja's view that each of these realms intertwines to produce and mediate our experience of urban landscapes. Case study 6.1 elaborates this relationship further by looking at the relationship between the city and the car. The growth of car ownership and use has had a profound impact on urban life. The car played an important symbolic and practical role in the development of industrial capitalism and consumer culture, and has been a central feature in the planning and building of cities. However, while the car generally plays a seemingly mundane role in people's everyday lives, there are rich consumption and representational cultures that directly relate car driving to urban life. The remainder of the chapter looks in more detail at the role of representation in both the development of the physical city and also how we experience urban life.

## Case study 6.1 **Consuming the city's streets – in cars**

Chapter 2 described the often mythologised and romanticised figure of the flâneur. This individual could experience the marvels of the architecture and street culture of the early modern city by looking up and down the boulevard. The flâneur could

continued

observe passers-by, enjoying the metropolis as a spectacle. It was in this milieu that the flâneur could move around the street, an anonymous middle-class observer, enjoying the urban sights, objects and people but never being drawn into meaningful social interaction. As we saw earlier, consumer culture emerges from this production of public urban spectacle and an urban world where modernity was a consumable experience.

The image and metaphor of the flâneur can be usefully updated to consider contemporary consumption of the city through the lens of the car driver. Sheller and Urry (2000), for example, argue that automobiles have fundamentally reconfigured the nature of urban social life in the twentieth century and produced distinct ways of dwelling, travelling and socialising in and through automobilised time-space. It can be clearly seen that there are parallels to be drawn between the flâneur of the early modern city (a detached observer, a pleasure-seeking stroller on the streets, a loiterer frittering away time) and the car driver. The flâneur's guiding principle is to look, but don't touch, strolling through the public spaces of the nineteenth-century metropolis. Car drivers move untouched through the city, as observer, and only carjacking, kidnapping, windscreen cleaners at traffic lights, crashes, horns, swearing, road rage and police intervention offer possible interruptions – in the same way that the mythical flâneur would have found it difficult to move around the city bereft from social context and without impacting or affecting urban social relations and forms of sociability.

Hall (2003) looks at mobility in the city and everyday life from a different perspective. While the most popular way of transport is the private car, he argues that cars unarguably have a negative impact on the local environment, non-renewable resources and urban societies and cultures. Cars are responsible for toxic emissions, noise, vibration and road accidents. Roads, and in particular motorways, are often responsible for a loss of public space, they sever communities or destroy sites of historic or environmental importance. The role of the car in refiguring urban life cannot be overestimated. The close relationship between urban development and the car is profound and can be seen as central to the development of shopping centres and drive-through restaurants, as well as the car itself becoming a space where food, drink, music and other commodities can be consumed.

Moreover, Hall highlights how the negative impacts of private transport are socially mediated. This is important because not all groups in the city suffer the negative impacts of private transport equally. The negative impacts of private mobility fall most heavily on the more disadvantaged groups in cities. These groups generally have the lowest rates of car ownership and restricted mobility. The more advantaged are able to insulate themselves from at least local impacts, by driving large, safe (from a driver's perspective) cars and living away from the areas blighted by heavy traffic and the negative consequences of road provision. Middle-class car drivers can pass through the marginalised and polluted parts of the city without encountering socially intense interaction with an urban 'other'.

Conversely, however, David Bell (2001) argues that there is both formal and informal sociality involved in driving a car. He suggests that this begins with the very privilege of passing a driving test, securing a licence and having to purchase other official seals of approval (for example, a tax disc and a roadworthiness certificate in the UK). Car drivers are also involved in a sense of communality, constructed through formalised conventions such as the highway code, as well as less formal concerns about polite and considerate driving etiquette. Bell also points to organisations such as owners' clubs, as well as informal chats about cars, their performance, style or colour as being commonplace. There are, moreover, trans-gressive activities associated with car driving, not solely related to legislative rules but also to inappropriate honking of the horn, degrees of road rage, picking up hitch-hikers, or picking your nose. There are also a raft of activities associated with socially responsible car use such as car sharing, and speed reduction in rural and built-up urban areas.

The car has reconfigured the social life of cities in a number of important ways. For example, the car has had a major impact on urban form, and cities have become more decentralised and dispersed, with the car essentially facilitating suburbanisation and urban sprawling. The environmental and economic impacts notwithstanding, there are also major social impacts of increased private transport. Hall argues that there are important issues of social exclusion in terms of individuals' lack of access to and hence participation in civic resources and opportunities. He argues that social exclusion has been exacerbated by the increased domination of the automobile of the planning and landscape of the city. Moreover, urban life is also qualitatively degraded by traffic and it has been shown that social interaction on streets declines as traffic volume increases. Where the car dominates our cities it is often to the detriment of human interaction and the marginalisation of the social and cultural life of urban areas.

However, despite the obvious negative consequences for urban life and issues of social exclusion that are related to car use, it should not be assumed that car driving is not a rich and diverse consumption experience in itself. For example, Edensor (2003) describes his daily drive as rich in mundane comfort and sensation, replete with small pleasures and diverting incidents and thoughts. He argues that although

continued

obviously environmentally and socially divisive, car driving is not necessarily an alienating urban experience.

Edensor shows that routine journeys can foster familiarity, and homely comfort, provoking affective and imaginative connections to times and places. He describes his car journey to work as a sensual experience of place, where memories, sensations, desires, fantasies and stories can be experienced in a homely micro-environment. Car journeys are not an empty experience, but social and cultural attachments can be formed as the driver's sensations over time generate varied emotional geographies and familiar dispositions towards the daily succession of spaces, places and landscapes. He shows how routine driving generates a flow of experience folding together – a fluid boundary of individual and collective experience that meshes together socio-cultural structures of feeling – in many of the same ways as other forms of consumption noted throughout this book. Moreover, what this case study shows is that the activity of car driving has had a profound impact on the physical development of the city, how use of the car is represented so centrally to our understanding of what urban life is about, and how the proliferation of the use of the car has impacted both negatively and positively on urban cultures. However, it is only by addressing these realms in complementarity that we can fully appreciate the complex relationship between people, cars and the city.

## (Re)presenting the city: place promotion

Maps, planning documents and plans, and place-promotional strategies are some examples of official representations of cities. This section and the following one look in more detail at Lefebvre's and Soja's consideration of the symbolic and physical representation of space. This section focuses on place promotion, which is one of the most interesting features of postmodern urbanity. The transformation of the image of the city and attempts to transcend the negative images that had become associated with the city following the de-industrialisation of urban economies have become more or less ubiquitous. All cities have an image – in fact, most have a number of images (Hall 1998). Place-promotion campaigns, which are local-government-sponsored advertising campaigns, can be characterised as simplified, generalised and often stereotypical representations of cities. Pursued through adverts in newspapers, magazines and trade journals, through advertising billboards, through glossy publications, and through festivals, carnivals and sporting events, place promotions seek to give cities a raised profile in order to attract investment, employers, tourists and new residents (Gold and Ward 1994).

Hall (1998) argues that it is impossible to know cities in their entirety, and place-promotion campaigns can only ever be a selective impression of complex reality. Place promotion typically exaggerates certain features, such as physical attributes, social diversity, cultural provision and infrastructure, favourable political conditions and economic competitiveness. However, Hall argues that these can often have little real resonance with the actual place; but this is not the point, rather it is the perception of images that is important. Since the 1980s there has been a growing industry around manipulation and promotion of place images. Cities attempt to overcome their bad image and the stigma that surrounds them as being poor and crime-ridden, having poor-quality architecture, a lack of job opportunities and second-rate cultural facilities. It is as an attempt to positively portray all these attributes that place promotion has become an integral part of urban regeneration. Nevertheless, as Table 6.1 shows, there are other ways in which the promotion and reputation of cities are communicated, beyond official and politically sanctioned imaging.

Hall (1998), however, suggests that place promotion is not a new phenomenon and has long been an integral part of urban development in a wide range of urban environments in the UK, North America, Europe and Asia. Some of the earliest examples of place promotion were during the nineteenth century with advertising campaigns attempting to attract people to the underpopulated American west. Municipal boards of trade, town councils, chambers of commerce and railway companies funded adverts in business and trade directories and local newspapers. However, over the course of the twentieth century progressively higher levels of state, provincial and federal government involvement in regional development programmes led to increased budgets for place-promotion activities in the US

### Table 6.1 *Formal and informal place promotion*

- Local authority programmes
- Media coverage of events which become prevailing impressions of those place (riots in various UK inner cities in the 1980s and in Los Angeles in 1982)
- Satire (comedians such as Billy Connolly and television comedies such as Rab C. Nesbitt have long used Glasgow's reputation for drunkenness and violence as a staple of their comedy)
- Personal experience (visits to cities; for example, tourist visits are frequently of short duration and by necessity highly selective, focusing on sites of interest and excluding large parts of cities)
- Hearsay and reputation (what people tell us about cities, whether from personal experience or hearsay, forms an important component of our impressions of place)

Source: Hall (1998: 120)

(Hall 1998). In the UK and Australia, place promotion was initially focused on different urban environments and populations. For example, the growth in nineteenth-century mass tourism in the UK led to a huge influx of holidaymakers to resorts such as Blackpool, Brighton and Margate, and was in part fuelled by railway companies eager to drum up business. Similarly, there was a concerted campaign by building societies to attract people out of London, to move to the growing residential suburbs.

While there was a differential proliferation of place-promotion campaigns throughout the western world at different times and with different foci, it is clear that place promotion has become more important since the 1970s (Ward 1994). This increased imperative has been stimulated by changes that have occurred in the organisation of the global economy. Place promotion has become increasingly important to cities hoping to maintain or enhance their economic competitiveness. Cities around the world are now part of a global urban system characterised by intense competition for investment and tourists. As cities underwent massive de-industrialisation, unemployment, and physical decline of their economies and urban fabric, place promotion has been used to market the attempts by cities to move (more or less successfully) to a post-industrial economy (Barke and Harrop 1994).

However, Hall (1998) shows that place-promotion campaigns have become more diverse than their predecessors. Rather than produce a unitary image of the city, the place-promotion campaigns of former industrial cities do not attempt to cater for a single homogeneous audience, but rather attempt to produce images that are attractive to a plethora of distinctive niche markets. Consequently, place-promotion campaigns tend to portray a variety of images. These include depictions of buildings and office space that are attractive to service-sector businesses, in an attempt to ensure a skilled workforce of largely middle-class professionals. Other images are oriented towards attracting business tourism, through representations of internationally renowned conference facilities and a thriving nightlife. The depicting of glass-fronted (often high-rise) office blocks seeks to appeal to middle- or higher-ranking representatives of companies, augmented again by images of high-cultural activity or accessible countryside. While such ubiquitous images are produced by many cities throughout the world, each place-promotion campaign still tries to represent unique and different features in order to secure a competitive advantage.

Images have assumed ever-greater importance in the post-industrial economy, and the actual production of urban landscapes needs to reflect the positive images (Gold 1994). As cities try to portray themselves as being economically competitive, the growth in spending as a proportion of local authority budgets has only been outstripped by spectacular and often speculative urban regeneration projects undertaken by city authorities. Local authorities have been involved with private companies in building mixed-use developments that create the type of business, retail or cultural environments so often used in place-promotion campaigns or projects to enhance the cultural lives of cities (see Chapter 7).

Place promotion has a number of key ways of spreading positive images. These can include the distribution of guides, brochures and other information through tourist offices, libraries and commercial information services. Other strategies include postal campaigns, poster adverts (for example at railway stations and airports) and press adverts in the financial and property pages of newspapers or in specialist property or commercial pull-outs (Myers-Jones and Brooker-Gross 1994). One of the most popular strategies has been through the employment of recognisable slogans and city logos (see Table 6.2). Cities also now frequently send representatives from local authorities and the business community to trade fairs and missions abroad. The twinning of cities in multiple countries on different continents throughout the world remains a widely used method of establishing economic and cultural links and exchanges.

A cursory survey of promotional literature reveals that almost every town or city promotes itself as a good place to work, play and visit. However, while the place-promotion slogans in Table 6.2 shows how cities emphasise not only their opportunities for business but also their lifestyle activities, over the past twenty years there has been a shift from an emphasis on the productive to the consumptive attractions of cities. Nevertheless, Hall (1998) shows there are several detailed images that cities have sought to create for themselves and promote as part of their economic regeneration. First is centrality, in that cities portray themselves as being at the centre of 'something'. This can include geographical centrality, making them more accessible than their competitors, suggesting easy communication and proximity to key markets (and hence low costs). The other important issue around centrality is cultural excellence, as cities seek to locate themselves as being endowed with an abundance of cultural activity, with bars, restaurants, nightclubs, theatres, ballet, music, sport and scenery.

The second key characteristic of place-promotion strategies is positive images of industry. Since the 1980s, former industrial cities have been amongst the most vigorous place promoters, in order to try to overcome widely held images of heavy, dirty, dangerous, polluting, rough, working-class activities. Images of the landscapes promoted are devoid of chimney stacks or steelworks but are full of highly designed green landscapes, science parks and shopping centres. The images that promote these features and the kinds of buildings that populate successful post-industrial activity are postmodern – ornate futuristic buildings populated

**Table 6.2** *Examples of place-promotion slogans*

Glasgow's Miles Better
Glasgow's Alive
Glasgow's on the Move
I ❤ NY
Skegness: It's So Bracing
Telford: A Better Way to Live and Work
Preston Multipli*city*
Alabama: The Right Place at the Right Time
Irvine: Always the Right Answer
Bedfordshire: Beautifully Connected
Blackburn: The Place with a Past and a Future
Cotton on to Burnley
Dudley: Tradition in Progress
Hull: The Gateway to Europe
Chesterfield: The Place to Grow
Smile: You're in Alloa
Sunrise Florida: The Center of Attention
Denver: Take a Closer Look
Santa Maria, California: A Liveable Place to Work
Tulsa: Out to Change your State of Mind
Wales: The Big Country
London Docklands: Dedicated to the Art of Work
Discover Islington: The Real London
Connecticut: America's Richest Cultural Life
Richmond Down: Where the South Begins
Oshawa: The Manchester of Canada
Stratford: The Hub of Rich and Fertile Western Ontario
Warwick: As a Trading, Residential and Educational Centre
Derby: Bounding with Vital Industries
South Tyneside: Catherine Cookson Country
Swansea: Dylan Thomas Country
Doncaster: England's Northern Jewel
Bedford: The Centre of England
Barnsley: Our Location is Our Excellence
Well Connected Wellingborough
Stoke-on-Trent: The City That Fires the Imagination

by light, high-tech and professional, managerial and business activity. Where impressions of manufacturing are represented then it is overwhelmingly images of technology, human skill, precision, cleanliness and healthy green environments. Such images seek simply to replace traditional negative images of industry with more positive appealing ones.

Hall identifies that one important element in marginalising traditional images of heavy industry has been the heritisation of former industrial sites as part of strategies to increase tourism. Large amounts of derelict or disused central city property, canals or docks have been restored, creating a desirable landscape of heritage sites to be used to generate positive images of places. Old industrial buildings are renovated to provide the kind of loft living noted in previous chapters, as well as providing workshop spaces for postFordist businesses. However, one of the most potent methods to suggest the marginalisation of an industrial urban past has been the growth in tourist attractions such as museums and visitor centres. Thus, while industrial tourist attractions have been criticised because of the heavily idealised, sanitised treatment of the past, depicting work and workers as cleaner and healthier than they actually were, the restored mills, docks, canals and heritage centres have become staple iconography for cities attempting to point to the fact that they have left their industrial past behind them.

The other side to this story of economic restructuring is the use of images of the urban business environment as dynamic places. Good business locations are central to modern urban promotion. Rather than mundane functional attributes of locations for business, place-promotion campaigns are often focused on images of architecture, communication and technology. This includes the representation of the spaces where business is conducted, such as office towers, convention centres and business parks – all crucial to the projection of a post-industrial city image and status. Buildings are icons of prestige, and business and architectural settings are vital in defining city status. Above all, the presence of skyscrapers and financial districts identifies an urban business culture as a nodal point helping to shape the global economy, in attracting investment and the relocation of progressive businesses with international aspirations.

The further core component of ubiquitous place promotion is images of lifestyle which sell cities as attractive places to live. Leisure time is considered increasingly important in decision-making for business relocation and is of course central to tourists' decisions to visit cities. However, there is often a narrowly prescribed notion of urban culture that cities have used in their promotional strategies. Overwhelmingly it is 'high' culture, such as theatres, ballet, classical music, art galleries and museums, that is represented in place-promotion campaigns. Such images are chosen to appeal to affluent middle-class professionals, and can also

**Figure 6.4 Heritage in the city. (Courtesy of the Black Country Living Museum)**

encompass exclusive wine bars, restaurants, designer shopping and nightclubs, as well as an abundance of exclusive and refined leisure facilities, particularly gyms, country clubs, spas and golf clubs. A final oft-utilised aspect of place-promotional strategies is positive images of an urban environment that is not dirty, overcrowded and unfriendly, but constructed around elements such as impressive architecture, green, clean and safe suburbs, and nearby easily accessible countryside.

In reflecting on the success of place promotion it is difficult to generalise about the success or influence of such campaigns in affecting relocation and investment decisions. Young and Lever (1997), for example, have undertaken empirical research to show that place-promotion strategies have little impact on the decisions of businesses thinking about relocating their premises. Despite the clear emphasis placed on such imagery in place promotion, evidence suggests that image is not so important. There has been no real attempt to gauge how a range of other organisations and individuals consume place-promotion strategies or the impact such images have on audiences beyond the business community. Nevertheless, despite little assessment of the actual impact of place promotion, the method is being widely used as part of a strategy for the economic survival of cities. Competition between places for tourism and business tourism may, as Jansson (2003) suggests, have led to an increased standardisation of images. Some would argue that place-promotion campaigns do nothing more than crudely and only partially mask the reality of problems facing many cities and space and places within them (Harvey 1989b).

It is clear, then, that place-promotion images only depict a partial representation of urban life. However, in their partial representation, they help us to elaborate on how institutional visions of the city, depicted by Lefebvre (representations of space and representational space) and Soja (as spatial practice and representations of space), are intertwined with the development of the physical city. This relationship in turn also highlights how changes in the physical fabric of the city and associated place-promotion images are consumed by different social groups, who are central to, or marginalised by, these images. The following section looks at a different form of representation – images of the city in popular culture – and argues that while popular cultural representations of the city are predominantly fictional and imaginary, they can be seen to have had a profound impact on urban life.

## The city and popular culture

The city has traditionally been a rich source of inspiration, and appears in many different popular cultural forms. The city regularly provides stories to fill the pages of newspapers and novels and it continues to provides topics, material and experiences for artists, film and television, musicians, advertisers, comics and storytellers (particularly in the form of urban myths). This section focuses predominantly on just one of those cultural forms, looking at urban images at the movies. The representation of the reality of the metropolis and how the city has a place in our imaginations will be addressed.

Before looking at movies, however, it is important to take a slight detour, and investigate the importance of the places where we usually go to see them – cinemas.

Donald (1999), for example, highlights the importance of cinemas to urban development. He argues that prior to the cinema the most prominent way in which the rapidly unfolding growth of urban life had been widely represented was by the popular press. Newspapers packaged a view of the urban world and made sense for their readers of the hurried and often anxious experience of the metropolitan landscape. Donald argues that newspapers offered two important things to urban dwellers: first, articles acted as a practical guide to surviving, exploiting and enjoying the city. And secondly, they depicted a collage of fragmentary stories of the city to be consumed by people actually reading them within the city – distractedly at home, in the workplace or on the tram or train between them. Donald suggests that following on from (and ultimately having a greater impact than) the popular press, cinema became the most important medium to portray urban life. Moreover, cinema was important because its growth was initially a thoroughly urban phenomenon in Europe and the US.

Donald suggests that while initially in the US the popularity of cinema was founded in working-class and immigrant ghettos, by the mid-1930s the explosion in the number of cinemas took them not only to central entertainment districts but also into the new suburbs as mass transportation made travel easier. Donald argues that along with national chains of department and grocery stores, cinemas quickly learned the power of marketing. Importantly, cinemas helped to give a sense of place to growing suburban agglomeration, and the location of cinemas was carefully chosen in order to attract maximum possible audiences. The golden rule for the locating of urban cinemas was to ensure that no potential patron needed to travel for more than half an hour to reach a theatre by car, train, tram or bus. It was also realised that once cinema-goers were in the cinema it was important to offer them an 'experience' and not just the film. Movie theatres were thus built with luxury surroundings, air conditioning and even babysitting facilities. Donald thus argues that in the emergent geography of twentieth-century urbanism, the cinema plays an important role. In one sense, cinemas helped to sustain and reinvigorate entertainment districts in the city centre and also helped to consolidate the suburbs. Cinema-going became a popular suburban pastime by bringing the experience of 'going out' to a way of life primarily built around 'staying in' – a way of life mediated increasingly through a privatised experience of telephone, radio and television.

However, Phil Hubbard (2002) revisits this argument during a trip to the multiplex. Hubbard investigates the building of multiplex cinemas in out-of-town locations during the 1980s and 1990s, and focuses on the sociality played out in these 'expolitan' leisure experiences. Hubbard argues that in contrast to city-centre cinemas, the growth in popularity of multiplexes can be explained by the way that they offer a predictable urban leisure experience, and provide the cinema-goer with an experience of riskless risk, by holding different social groups and socialities apart.

Hubbard uses the case study of Leicester, in England's Midlands, and shows that expenditure at the cinema increased threefold in the 1990s. He argues that this increase is underpinned by the proliferation of family entertainment centres that changed the geography of film. Hubbard is interested in understanding the ways in which the development of these new spaces of film exhibition has impacted upon consumption cultures and urban life. He argues that the multiplex has turned its back on the city, and generates a form of leisure that is placeless. However, while the building of multiplexes has increased cinema-going in the UK, it is people between 16 and 35 years of age who are most drawn to multiplexes because of their predictable and safe sociality. Hubbard (2002) suggests that the appeal of multiplexes for such groups is that they allow risk avoidance, thus reinforcing the boundaries between self and other. Hubbard suggests that in a context where fear saturates our everyday lives, mundane acts are now viewed as risky and dangerous. Of course, some activities and some places (such as city centres) are regarded as more risky than others, and multiplexes are perceived to be low-risk and safe family spaces. However, the sociality and thus the popularity of multiplexes in essence rely on the exclusion of certain groups (typically the less affluent and non-white). Hubbard concludes that going to multiplexes may not be very different from staying in to watch a film on television or on video or DVD.

The importance of the cinema to urban life is not, however, restricted to its role in facilitating mass consumption, but comes from the way in which films play an important role in representing and imagining urban life The cinema suggested an informed and participatory urban public – a community of spectators being educated for modernity and its concern for progress and technology. Moreover, as Clarke (1997) shows, the cinema became an urban space that offers ways of imagining the city. Table 6.3 shows just a small number of examples of the large number of films that are focused on urban contexts. Such films as *King Kong* and genres such as *film noir*, urban horror and suburban comedy show that the city has featured in many different ways as central to the plot of a wide variety of films, with different themes and messages (Donald 1999). While the films represented in Table 6.3 vary in terms of genre, narrative and a variety of tropes that can be grouped together in order to represent discourses about the city, each offers in its own right a particular vision of urban life that can also have economic and cultural consequences for the 'real' city (often in specific cities). Amelie, for example, represents an image of Paris that has become highly lucrative as tourist product; Reykjavik 101 has sedimented certain myths about Iceland as being cool, hip and happening; Notting Hill seals a myth of white consumerist London, and so on.

## Table 6.3 *The city at the movies*

| | | |
|---|---|---|
| The Italian Job | Lock, Stock and Two | When Strangers Marry |
| Escape from New York | Smoking Barrels | Breakfast at Tiffany's |
| Seven | 51st State | The Big Sleep |
| Mary Poppins | Quadrophenia | Blue Velvet |
| My Fair Lady | Suburbia | LA Confidential |
| Pretty Woman | ET | Alphaville |
| Last Tango in Paris | Fatal Attraction | Halloween |
| Rosemary's Baby | Poltergeist | Night on Earth |
| 24 Hour Party People | Wall Street | Nineteen Eighty-Four |
| Amelie | Disclosure | Just Another Girl on the |
| Godzilla | Taxi Driver | I.R.T. |
| The Day after Tomorrow | Human Traffic | Mean Streets |
| King Kong | Blade Runner | Manhattan |
| Midnight Cowboy | Metropolis | Tootsie |
| Death Wish | Predator 2 | Mrs Doubtfire |
| Beverly Hills Cop | Robocop | Flash Gordon |
| La Haine | Local Hero | Who Framed Roger |
| Maid in Manhattan | Sunrise | Rabbit |
| Reykjavik 101 | Mr Deeds Goes to Town | Dr Jekyll and Mr Hyde |
| Candyman | Sleeping with the Enemy | Demolition Man |
| Blade Runner | Single White Female | Notting Hill |
| Die Hard | Assault on Precinct 13 | City of Angels |
| Speed | Big Trouble in Little | Wings of Desire |
| Trainspotting | China | Manhattan |
| Falling Down | Witness | Sleepless in Seattle |
| Strange Days | Death Wish | Paris When it Sizzles |
| Boyz n the Hood | Singin' in the Rain | Saturday Night and |
| Superman | Vertigo | Sunday Morning |
| Batman | Cabaret | Up the Junction |
| Spiderman | An American Werewolf in | Things To Do In |
| Ghostbusters | London | Denver When |
| The Full Monty | My Beautiful Laundrette | You're Dead |
| Gangs of New York | Mr Smith Goes to | Man Bites Dog |
| Leaving Las Vegas | Washington | Gregory's Girl |

Such films inform and mediate our knowledge, understanding and experience of urban life. In everyday experiences cities frequently seem to possess a cinematic quality, and urban studies has only just come to think about this relationship. When we visit cities they can seem to have stepped right out of the movies – the cityscape has an intimate relationship to the screenscape (Clarke 1997). Indeed, the city in film is not only a useful tool for attracting tourists; there is also a healthy competition amongst cities around the world that seeks to attract film-makers to use the 'real' city as locations for films. Yet despite the obvious and immediately perceptible cinematic qualities that cities frequently seem to possess, and despite the uncredited role played by cities in so many films, relatively little academic attention to, and understanding of, the relationship between urban and cinematic space has been published. While there has been an implicit acceptance of the importance that cities have in popular cultural forms – in particular, visual representation of the city in paintings and photography – there has been relatively little explicit attention to theorising the urban themes and storylines encompassed within films (Clarke 1997).

David Clarke's (1997) comments notwithstanding, the role played by the cinema in rapid urbanisation and the expansion of industrial capitalism, and thus the ways in which cinema helped shape the historical transition towards specifically modern modes of living, have been acknowledged in theoretical and empirical studies (Donald 1999; Hubbard 2002). However, it is clear that not only did cinema document and provide commentary on such developments, but it was a cultural form that was implicated in contributing to these changes in other ways. For example, the spectacle of cinema both drew on, and contributed to, the increased pace of modern city life. Cinema helped to normalise the newness of the frantic sights and sounds of the modern city. Cinema reflected and helped mould the novel forms of social relations that developed in the crowded yet anonymous streets. Cinema both documents and helps normalise and imagine the transformation of the social and physical spaces of the modern city. Cinema has helped to instil urbanism and consumerism as a way of life. For example, futuristic representations of cities in films can be seen to have a great impact on the way we all image the possibilities and failings of urban life. Such images inform not only planning, architecture and urban design, but also everyday imaginings and dreams about the possibilities of future urban life, good or bad.

However, Clarke (1997) elaborates on this point by arguing that while it is right to investigate the visual element of the cinema, our experience of movies cannot be reduced to one simple reading. Clarke argues that in achieving a reframing of the city, the camera's penetration of reality entails a transformation in the perception of cinema-goers, but does so in a way that is consonant with the viewers' knowledge and experience of the city. Thus, it must be remembered and stressed that there can never be one single interpretation and representation of the city. For some, for instance, the concept of the stranger and their physical proximity yet social distance led to the perception of a world populated by strangers. This contributed to an imaginary urban sphere of people strolling around the city with an ambivalence to their fellow streetwalkers – modernity's unintended consequence was a surprising new form of ambivalence. Conversely, in the postmodern city, lifestyles are increasingly geared to pleasures and sensations without consequence. Both conceptions of urban life, as we saw in the previous chapter, are very much determined by the enabling and constraining nature of social construction of class, gender, ethnicity, sexuality, and so on.

This point is explained by Stevenson (2003), who argues that cities are understood and experienced in a range of contradictory yet reinforcing ways. Fundamental to this understanding is the interplay between the 'real' city of lived personal experience and the 'imaginary' city of representation and fantasy. The tangible city is made up of footpaths, buildings and roads, and the imagined city is the place in literature, popular culture, anecdote and memory. Of course, the city of experience is also the city of dreams and nightmares; it is a city of friendship, loneliness, fear, avoidance, memory, love and home (Stevenson 2003). The imagined city thus intersects with the real to construct intimate personal relationships with place.

It is in this way that Stevenson (2003) shows us that representations of the city can serve as anchors that confirm we are (or have been) here/there/somewhere/ anywhere. It is possible to locate oneself in space using the coordinates on the map; we recognise and relate to the buildings and places we have gone to and those we have only ever seen in movies or read about – this is how the urban is framed in the imagination. Stevenson argues that throughout the twentieth century, the representation of urban form became increasingly global and immediate. The development of international circuits of information and communication, most significantly in the production and consumption of film and media images, makes it possible for selected representations of urban landscapes and urban life to be conveyed routinely to audiences around the globe.

For instance, popular culture provides people with access to cities they have never visited and thus with a strong impression of their physical and symbolic form – Central Park, skyscrapers and wastelands, elevated railways, traffic jams. The futuristic imagery of the Los Angeles of *Blade Runner* is decaying and dehumanising, while Fritz Lang's classic 1920s film *Metropolis* remains a powerful metaphor for a class-divided society. Stevenson (2003) argues that few encounters with the city and the complexity of living in cities can be more evocative or confronting than those experiences through popular cultural forms such as literature, art and television. She suggests that we come across so many cities in popular culture – from the city of fear, drudgery and loneliness to the city of utopian experience. In the virtual world of popular consumer culture we can encounter entire cities (or parts of them) that we may never have visited, as well as those we believe we know well.

In all genres and national locations, film, literature and the popular media remain influential sources of images and ideas of urbanism and the urban landscape.

**Figure 6.5** *An image from the film* **Metropolis.** *(Courtesy of Eureka Entertainment)*

Representations of cities are intriguing markers of a whole range of diverse imagined urban cultures. There are images of the city as wilderness – untamed, dangerous, and inhabited by wild animals, a place of base instincts, ugly motives – in films such as *Escape from New York* and *Falling Down* (see Case study 6.2). Another important cinematic trope is to explore the relationship between the city and suburbia (*Fatal Attraction*, *ET*, *Poltergeist*), detailing both the community and utopian visions of suburbia as well as its more dysfunctional concerns. Similarly, films like *Boyz n the Hood* highlight the poverty, crime and drug problems related to ghetto life.

## Case study 6.2 *Falling Down*

Malcolm Miles (2004) argues that for many who watched the destruction of the World Trade Center, New York, on 11 September 2001, it would have felt as though they were watching a disaster scenario often witnessed in films. The

continued

images of the collapsing twin towers, the twisted steel frame, dust clouds and the sickly smoke still rising two months on from what was in part a pile of human ash were for Miles a further chapter to add to dystopian urban imaginaries. Such urban catastrophe has long been a favourite of film-makers as well as in popular fiction. There are many such urban disaster films, as seen in the list that makes up Table 6.3.

However, this case study focuses on *Falling Down* (1992), a film that has many interesting elements of debate and concern about contemporary urban life in the USA. In *Falling Down*, Michael Douglas plays William Foster, a man dislocated from the city of Los Angeles, encoded as alienated and fragmented and, for Douglas's protagonist, a city to play out a male white-collar crisis of masculinity. William Foster is just trying to get to his rented apartment through the 'urban jungle'; and his travels become a violent and alienating experience. Foster is an obsolete ex-family man in a dead-end job and is frustrated with his experience of urban life as hostile, violent, sexually deviant and unreadable – out of control. In *Falling Down*, the city is rotting, claustrophobic, invasive and stagnant, characterised by an endless wait, going nowhere in a seemingly endless traffic queue. Foster feels no sense of ownership or rootedness in the city, but conversely embodies a sense of entitlement to the city, and his encounters on his journey home traverse the socially and segregated city where he feels no real belonging. He encounters a city 'taken over' by racial/and or sexual 'others', and is ejected from the golf courses of gated communities, where he is also unwelcome.

Foster is displaced from the domestic sphere of his ex-wife and estranged children, and he is alienated from the public sphere that is violently dominated by ethnic or sexual deviants, while his white-collar status excludes him from a more affluent urban life. He tries unsuccessfully to board a bus that is surrounded by a large crowd, and he experiences public space as claustrophobic and unreadable – he feels ill at ease and unable to judge or culturally or socially understand or empathise with his surroundings and his fellow urbanites. In public space he is also confronted by the urban 'other': carnivalesque groups of buskers and beggars (one holding a sign that reads 'We are dying of Aids'). He witnesses a fight between people on the street, and the music, colour and heat assault his senses.

It is not only public space from which Foster is dispossessed and obsolete. The narrative also deals with other spaces to mark his alienation. For example, the cityscape is contrasted throughout with the seemingly safe maternal, feminised space of the 'home'. The use of this juxtaposition as a structuring, narrative device is particularly striking. The protagonist descends into violence and out-of-control masculinity as he struggles with a gang seeking to protect their territory. But for all the violence and extremity we are supposed to find Douglas's character empathetic; if not his violent reaction to his lived experience. The use of slow motion and apocalyptic music seeks to depict the city as unreal. This helps to frame the story-line that attempts to depict and explain why Foster (a white-collar alienated worker) turns to violence in his attempts to cope with the corrupt and crazed world of the contemporary US city. This vision, while being less dramatic than other apocalyptic and disaster scenarios, is nonetheless dystopian: a vision of the city with social conflict, decay, violence and loss of hope.

Source: Mahony (1997)

Stevenson (2003) argues that there is a complex relationship between the real and the imagined city – between the city of everyday life and political and economic institutions, and that of film, television and other forms of cultural expression. Nevertheless, representations of the city can reveal insights into social relations, social and cultural concerns, and ideas and fantasy. Similarly, representations of the 'official' city in maps or place promotions do not deal with the real lived experience of cities and are always partial, selective representations of spaces, realities and spatial relationships, but nonetheless tell us a lot about the vision of urban life, and the economic and political priorities of politicians, planners, architects, and so on. Popular cultural forms have vigorously represented the city in many different ways, but the way we all consume them relates to different social relations and identity positions and the different places that people live in the world.

## Conclusion

This chapter has investigated the relationship between the physical city, the 'official' city as represented in maps, place promotion and planning, and the city as represented in popular cultural forms such as film, adverts, television and music, and finally how people consume the city in everyday life. It has been argued that these three realms are folded together, intimately informing each other. Each has its impact on the others and all are mutually dependent. The physical city, for example, is affected both by planning and by the fantasy and imagination of popular culture. Everyday urban life is both impacted upon and informs official and popular representations of the city. Our interpretation and exposure to popular cultural representations of urban life can have a profound impact on how we view the city, places within it, its people and cultural activities, and so on.

An important argument throughout this chapter has been to show that the city is produced and consumed in multiple ways, and there is no one single representation, interpretation or experience that can be said to know, describe or encapsulate a city, spaces or places within it, or aspects of urban life. Our interpretation and experience of consuming the city, whether it is real or imagined, is very much dependent on social (and hence power) relations that surround constructions of class, gender, ethnicity, sexuality, age, nationality, and so forth. As Lefebvre and Soja show, the social production and consumption of the city are bound up in the interplay of three spatial realms – the real, imagined and experienced. Nevertheless, this chapter has highlighted a number of images, metaphors and discourses through which the city is represented, and the texts which both help to mediate our experience of the city in the present and influence the way it will develop in the future.

> **Learning outcomes**
>
> - **To have an understanding of the relationship between the physical city, representations of the city and everyday experiences of urban life**
> - **To be able to describe the ways in which official representations of the city seek to construct a particular image of physical, political, economic, social and cultural competitiveness**
> - **To offer insights into the diverse and varied representations of the city in popular cultures**
> - **To be able to critically appraise the relationship between the imagined and the 'real' city**

## Further reading

Edward Soja (1996) *Thirdspace: Journeys to Los Angeles and Other Real-and-Imagined Places*, Oxford: Blackwell. This book is a comprehensive primer to theories of the social construction of spatiality. Soja synthesises the work of theorists such as Lefebvre and Foucault as well as bodies of literature around feminist, postcolonial and postmodern thinking.

Anthony D. King (ed.) (1996) *Re-presenting the City: Ethnicity, Capital and Culture in the 21st Century Metropolis*, Basingstoke: Macmillan. An edited book that draws on a wide range of disciplines including geography, architecture, art history, sociology and politics to describe the construction of the real and imagined city.

David B. Clarke (ed.) (1997) *The Cinematic City*, London: Routledge. This edited book describes the roles played by the city in a wide variety of films. The book traverses urban and film theory and provides an interesting critique of a range of cinematic forms related to the city, and a wealth of empirical detail.

John R. Gold and Stephen V. Ward (eds) (1994) *Place Promotion: The Use of Publicity and Marketing to Sell Towns and Regions*, Chichester: John Wiley. This edited volume provides a comprehensive introduction to place promotion in different contexts from cities to regions to the countryside and in national contexts such as the UK, USA, Australia and Canada. Chapters provide theoretical debate and empirical evidence in a balanced and complementary way.

Nicholas R. Fyfe (ed.) (1998) *Images of the Street: Planning, Identity and Control in Public Space*, London: Routledge. A rich and interesting volume that describes the diverse nature of streets throughout the world. Chapters show how streets are sites of complex social relations and political protest, sites of domination and resistance and pleasure, danger and anxiety.

# 7 Consumption and urban regeneration

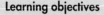

Learning objectives

- To look at the relationship between consumption and urban regeneration
- To describe varied strategies and outline examples of consumption-led urban regeneration
- To explore consumption cultures that are supported by local governments
- To think about the ability of cities lower down the urban hierarchy to utilise consumption-led regeneration in order to improve their competitiveness

This chapter elaborates the view that in a global urban hierarchy characterised by intense competition, cities are promoted and sold not simply as centres of economic growth but as culturally rich places in which to live and work, where the quality and quantity of consumption opportunities are crucial elements in generating competitive urban conditions. The chapter will argue that central to cities' attempts to move away from a dominance of industrial production to a post-industrial economy has been the development of political and economic activity that renders invisible the industrial past, as well as a cultural and creative economy which enhances a city's liveability. This chapter will critique the seemingly ubiquitous presence of consumption-led urban regeneration initiatives and their implications for cities throughout the urban hierarchy.

## Consumption-led urban regeneration

Chapters 2 and 3 introduced geographies of consumption that have been integral to the development of our cities. Arcades, department stores, central commercial districts, spectacular shopping malls, loft living, and so on became some of the most important spaces and places of urban consumption. This chapter looks at urban developments that have taken place over the past twenty-five years. There is a focus on urban regeneration projects initiated by local government authorities that have sought to overcome the problems of de-industrialisation that affected cities around the world during this period. It will be shown that cultural production and consumption have become key elements in such strategies, which are undertaken in order to enhance the reputation of cities via the range of consumption opportunities they generate.

Chapter 3 showed that since the 1980s the global restructuring of economic, political and social processes has had a profound effect upon the nature of everyday life. Related to this profound change has been the decline of manufacturing industries and an increase in the importance of business, professional and service industries – the simultaneous increase in mass unemployment, physical decline, and the rise of a 'new petite bourgeoisie' (Giddens 1973). As a result, increased competition has arisen between cities as they try to create new images in order to attract speculators, businesses and consumers. If cities were to be successfully rejuvenated, local authorities became aware that they would have to initiate rede-velopments which were often speculative in nature and high-profile, and had symbolic significance in attempting to represent progressive (rather than regressive) socio-cultural and economic trends. Such redevelopment is characterised by new public–private partnerships and programmes of place promotion, but is also visibly more spectacular, with revitalised city centres and agglomerated business and financial districts featuring gleaming high-rise office blocks, waterfront devel-opments, heritage centres and urban villages. Augmenting this has been the promotion of the creative industries (advertising, architecture, the art and antique market, crafts, design, designer fashion, film, interactive leisure software, music, performing art, publishing, software, television and radio). With buildings and facilities such as theatres, art galleries, parks, convention and exhibition centres (as well a supporting cast of café bars, restaurants, fashion boutiques, delicatessens and other cultural facilities), the buzz of 'creativity, innovation and entrepreneuri-alism' – assumed to be associated with these activities – is increasingly seen as crucial to the competitiveness of cities (see Hall and Hubbard 1998).

## Consumption and spectacular urban regeneration

As urban life has developed since the rapid growth of the modern city, most cities and towns boast that they have a range (if not all, in most cases) of important consumption spaces such as city-centre shopping districts, shopping arcades, department stores, out-of-town retail parks or even mega-malls. The size and reputation of such consumption opportunities, and their ability to attract people from the farthest possible distances, increase towards the top of the urban hierarchy – world cities retain their status in part due to the consumption activities that they can sustain.

Since the 1980s, in response to problems associated with de-industrialisation, local authorities have played an increased role in developing spaces of urban consumption. Such strategies have been vital if cities are to maintain or generate the reputation of possessing the most up-to-date consumption opportunities. Although these strategies are generally described in terms of supporting the 'cultural' life of the city, local authorities have had to support the consumption attractions of the city – in order to attract investment and post-industrial businesses, which in the new global economy could in theory locate their operations anywhere in the world. City authorities saw the strength of their cultural facilities and infrastructure and the consumption opportunities they offered as vital to attracting global capital and tourists. This has led to the almost totally ubiquitous situation that cities throughout the world, although with differing degrees of success (this will be returned to later), attempt such cultural- (or consumption-) led regeneration. What follows is a description of the consumption- and culture-led regeneration initiatives of just three – Birmingham, Singapore and Chicago.

*Birmingham* – Tim Hall (1997) describes the regeneration of Birmingham in the UK, a city in the English Midlands, as an attempt to overcome the marginalisation of the city in terms of the cultural geographies of England, which had popularly labelled the city as uncultured. One of the centrepieces of this strategy was the development of a convention centre, the idea for which was raised in 1981 at the height of economic recession in the UK. The convention centre idea was quickly taken up as a panacea to reverse the effects of de-industrialisation. Work began in 1986, funded by a European Regional Development Fund grant of £49.75m, and the International Convention Centre (ICC) was completed at a cost of £180m in 1991. The ICC consists of eleven halls of varying sizes, one of which, Symphony Hall, was specifically designed as an international concert venue (see Figure 7.1).

However, as Table 7.1 shows, in a brief and by no means comprehensive list of urban regeneration initiatives that have taken place in Birmingham since the 1980s,

**Figure 7.1 The International Convention Centre, Birmingham, UK. (Courtesy of Tim Hall) Source: Hall (1997)**

**Table 7.1 _Birmingham's flagship regeneration projects_**

- The £80m International Convention Centre (ICC), opened in April 1991, with a maximum conference capacity of 3,700 delegates and the inclusion of a 'world-class' symphony hall;
- The £60m National Indoor Arena (NIA), built to enhance the city's position as an international centre for sports and music, with seating for up to 12,000;
- The £31m Hyatt Hotel, built as an integral part of the ICC development;
- The £250m privately financed Brindley Place festival marketplace scheme. The scheme included 850,000 square feet of offices, 123,000 square feet of retail space and 143 houses;
- The Jewellery Quarter: the designation and marketing of an area historically associated with jewellery production. The quarter is home to small-scale jewellery makers, a museum and retail space;
- The Chinese Quarter is an area of the city that had organically grown around the agglomeration of Chinese restaurants and businesses, and was one element of regeneration that sought to give the city a cosmopolitan feel, and make the city legible to tourists and visitors who may have visited such quarters or 'Chinatowns' elsewhere in the world;
- The Custard Factory is a privately owned but publicly funded creative-industries-oriented project. This mixed-use development at the edge of the city centre combines business incubator space, shops, restaurants and bars, and provides a focus for both cultural production and consumption;
- The Eastside district of the city has been earmarked for the development of an urban park, small incubator units and further creative industries development initiatives;
- The Bullring shopping centre has been redeveloped with £5000m of both public and private funding. The development boasts retail space the size of 26 football pitches, with over 140 shops, boutiques and restaurants at the heart of the city.

Sources: Hall (1997), Chan (2003), Beazley _et al._ (1997), Jayne (2005)

the ICC was but one of a range of culture-led regeneration initiatives. Hall argues that the ICC was part of a process of making visible the invisible – the re-imagining of the city centre through the depiction of buoyant cultural life via its flagship projects (which also included the pedestrianisation of squares and plazas and the commissioning of public art). One of the most high-profile, shown in Figure 7.2, was the sculpture _Forward_, by artist Raymond Mason. The sculpture narrates the emergence of civic pride through the figurative representation of numerous local 'ordinary' people. While _Forward_ celebrates the city's industrial past, it acts as a monument set amongst those projects that sought to generate post-industrial urban economic and cultural activity. _Forward_ seeks to provide a memorial to a past life of the city, now left behind.

**Figure 7.2 Forward.** *(Courtesy of Tim Hall)*
*Source: Hall (1997)*

However, Beazley *et al.* (1997) are right to show that despite the massive investment that has been made in the city centre, disadvantaged groups not only have gained few benefits from the projects (which are ultimately aimed at attracting the money of those who are economically better off), but have also suffered from the diversion of scarce funds away from services such as housing renewal and education. The perceived image of Birmingham as being cultureless led to major (and effective) consumption- and culture-oriented spectacular developments in the city centre. These have been very successful in increasing Birmingham's profile as a location for conference-goers and business tourists. Nevertheless, following a change in the political regime, coupled with local popular disquiet, policies are now more directed (at least in paying lip-service) to implementing welfare and community-based policies than to pursuing further large-scale, expensive and potentially socially detrimental spectacular developments (Loftman and Nevin 1998).

In 2004, Birmingham's faith in spectacular regeneration was restored, however, with the redevelopment of one of its most infamous landmarks, the Bullring Shopping Centre (see Figure 7.3). Built through a £500m partnership between the public and private sectors, the new Bullring replaced its bleak 1960s predecessor, and at its heart has an iconographical Selfridges department store, amongst the new

**Figure 7.3a** *The newly developed Bullring Shopping centre, Birmingham, UK. (Photo: Mark Jayne)*

designer boutiques and restaurants. The new Bullring seeks to compete in an economy in which Birmingham was fast falling behind its rivals such as Manchester, Leeds and Glasgow. The Bullring has become the centrepiece of consumption-led regeneration in the city, and with its central location and focus on shopping is an important addition to regeneration activities that have been undertaken in the city over the past twenty-five years.

*Singapore* – Chang (2000) describes the attempt by a relatively competitive world city to become known worldwide as a centre for the arts. Already possessing an international air transport hub and known as a successful convention city, a branding as *Renaissance City* had been adopted by the city government as it attempted to import a western-style culture-and-entertainment economic growth strategy. The tag of becoming a *Global City of the Arts* sought to make Singapore an investment base for cultural businesses and enterprises in the region, in order to make the city into the theatre hub of South-East Asia, and further to portray it as a cosmopolitan destination for global tourists.

**Figure 7.3b** *The spectacular Selfridges department store at the newly developed Bullring Shopping centre, Birmingham, UK. (Photo: Mark Jayne)*

In order to achieve these goals, the local city authorities planned a major development, the *Esplanade-Theater* at the heart of Singapore's bay. This $250m project comprises an 18,000-seat concert hall and a 2,000-seat lyric theatre next to a modern Marina Bay hotel and retail complex. As Figure 7.4 shows, the consumption of cultural activities and buildings is clearly seen by the Singaporean authorities to be a coherent regeneration project for a former powerhouse industrial city. These major urban developments have combined foreign capital and local authority money, and have become focused on the centrepiece, the Shanghai Centre, a vast multi-use complex comprising 472 luxury apartments, 25,000 sq. metres of office space, a theatre, an exhibition atrium and a five-star hotel.

It is clear in the case of Singapore (as with Birmingham) that the culture-led urban regeneration undertaken in high-profile city-centre locations is driven by global competitive pressures. However, Chang (2000) argues that while such projects are underpinned by the need to participate in an international cultural marketplace and are an attempt to localise global stylistic trends, such developments generally

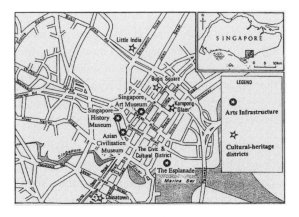

**Figure 7.4** *Singapore's central cultural regeneration. (Courtesy of T. C. Chang)*
**Source: Chang (2000)**

do little to promote, support or fund local cultural production. Moreover, Chang (2000) suggests that the international renown of such developments does not in turn generate a reputation effect that bestows a reflected profile on local artists (allowing them to gain recognition). He suggests that the programming of 'high-culture' and internationally oriented productions that dominate the flagship cultural projects built in the urban core of Singapore is in fact having the dual effect not only of redistributing funding away from local cultural production, but also of appealing to only a small minority of the local population.

In a similar vein, Bianchini and Ghilardi (2004) argue that such flagship projects generally fail to adopt a holistic approach to cultural provision and planning. The focus of city authorities on large-scale projects is often achieved at the expense of financial support of local cultural production and consumption cultures. Moreover, Graeme Evans (2001) argues that the substantial financial cost of flagship cultural projects belies the mismatch between such spending and local people benefiting from 'trickle-down' economics. Such global-facing culture-led regeneration projects are focused predominantly on globalised economic accumulation strategies at the expense of cultural inclusion and the support and representation of the diverse cultures that constitute urban life.

*Chicago* – Augmenting, and often included within, the kind of physical and symbolic attempts to improve the urban environment as seen in Birmingham, Singapore and many other cities around the world are initiatives to promote the 'creative city' (Landry 1995). This notion encompasses aesthetic improvements of 'soft infrastructure' ranging from the building of squares and fountains to the greening of streets, the provision of benches and improved public spaces, the establishment of late-night shopping and 'happy hours', cultural events and festivals such as music, literary or street theatre – all designed to make the city more 'liveable'.

For example, conceived in 1997 and championed by Chicago's mayor, Richard Daley, Millennium Park is a project on a large scale. It has taken shape on the edge of the city centre, partly superimposed over the city's Grant Park, which was expanded to form a 24-acre site, some of which sits on top of sunken railway tracks. The sheer scale and cost of the park are an attempt to reinforce Chicago's reputation as a centre of contemporary arts, as well as reinforcing the city's renown as being the cradle of modern office (skyscraper) architecture in America (Usborne 2004).

By the time of its opening in 2004 (and, as its name suggests, after a four-year delay), the park had cost over \$400m (£222m), twice as much as originally envisaged. The funding of the park was generated through public and private finance. A coalition of donors, spurred on by civic pride, contributed \$120m, mostly to pay for artworks in the park. No fewer than 48 individuals, 20 corporations and 12 foundations each granted \$1m or more to the park. Private benefactors include Oprah Winfrey, television stars and the retired head of the Sara Lee Corporation.

At the core of the park is a music pavilion designed by Frank Gehry, whose buildings around the world include the Guggenheim Museum in Bilbao. There is also a modernistic fountain, complete with video screens, by James Pleanesa, a Catalan designer. However, the centrepiece of the park is a sculpture by Anish Kapoor. As yet unnamed, the 125-tonne sculpture is 66 feet long, 32 feet tall and 47 feet wide. The sculpture has an instep that creates a space for the public to stand beneath, as if they were in a reflecting cave. They can look up to see their distorted reflections in its surface. This is Kapoor's first public piece of art in the US, but discussions are under way for other US cities to add cachet to their regeneration through such symbolic and artistic visions of their bright futures (Usborne 2004).

While the development of an urban park may not at first appearance be hindered by the exclusionary factor of the entry price that is levied at other flagship projects such as theatres, art galleries, museums, and so on, there remain critical parallels to be drawn. DeFilippis (2004), for example, describes how the contemporary production of urban space in US cities is inherently exclusionary and that large-scale redevelopments come at a price. He suggests that such projects are generally driven by aesthetic and architectural goals, financed by public–private partnerships (with a large burden being placed on public finds) and are surveillance-heavy. DeFilippis argues that participation in urban space in the US is defined by an institutional view of the citizen as consumer, and hence the production of spaces is oriented around middle-class consumption cultures. He suggests that such policy ultimately leads to an oppressive socio-spatial production of space. DeFilippis notes that US public space (including public parks) is characterised by intense class- and race-based exclusions and segregation that erase (and with the

**Figure 7.5a** *The Frank Gehry-designed Pavilion, Chicago Millennium Park. (Courtesy of Charlie Scheips)*

intervention of police and private security guards who physically eject) people who represent the urban 'other'.

It is clear from just these three examples of Birmingham, Singapore and Chicago that in this vision of the restructuring of cities – particularly those which have most successfully moved from manufacturing production to a service and consumption-based economy – that specific (social and spatial) forms of urban life are produced. Central to such strategies of urban regeneration are processes of rescripting what the city is about. This rescripting includes a conscious manipulation of the cultural representations of the city in place-promotion campaigns and the media. As was shown in Chapter 3, this is often centred around highlighting certain physical spaces/places in the city and the business, financial and professional employment opportunities, as well as the plethora of bars, restaurants and theatres which such social groups are thought to prefer (Jayne 2000). This can be best summarised as embodying an idealised white middle-class hegemonic notion of what cities should be like.

This political, economic, social and spatial process is inherent in a 'symbolic economy' represented not only in the physical spaces of the city (and in

**Figure 7.5b** *Anish Kapoor sculpture, Chicago Millennium Park.*
*(Courtesy of Charlie Scheips)*

place-promotion campaigns and official policy documents) but also in the virtual space of the consumer society, and in popular cultural forms (such as the mass media). However, what becomes clear from the examples of Birmingham, Singapore and Chicago is the growing significance of public–private partnerships in urban regeneration initiatives. A significant feature of these partnerships is that a corporate agenda is often driving urban change. The role of the public sector appears to be increasingly about facilitating development opportunities, cleaning up the streets, and ensuring the marginalisation of 'undesirable' people and activities. This takes place hand in hand with allowing public funding to generate urban spaces and places around an urban middle-class cappuccino culture, and local authorities hoping that such consumption takes root, and attracts further inward investment.

Increasingly, then, what distinguishes one place or one city from another is the strength of its consumptional identities – the range of shops, restaurants, museums and other tourist attractions, architecture, streets, and so on that it possesses. As places compete for limited investment funds, their vitality and viability depend on sustaining and nurturing a particular image. Central to such place-selling strategies is a conscious and deliberate manipulation of culture in an effort to enhance the appeal and interest of places. Places are thus being promoted and sold not simply as centres of economic growth but also as culturally rich places in which to live and work (see Chapter 6). Central to this is the quality and quantity of consumption opportunities, and hence consumption that becomes central to generating place myths. What is crucial here, then, is that contemporary consumption is intrinsically linked to quality of life (Crewe and Lowe 1995).

In older cities, such strategies emerged in urban regeneration programmes in the absence of specialist alternative business developments. However, cities that were dominant in the industrial urban hierarchy have tended to continue to be 'cultural [or consumption] capitals' and tourist centres in the post-industrial urban hierarchy. These positions appear to be maintained through the development of sustainable post-industrial business, professional and service sectors and a constant flux of innovation and competition in order to maintain competitive advantage. However, it must be stressed that, as with individuals and social groups, cities (and places/spaces within cities) are subject to the 'consuming paradox'. Urban change and regeneration projects, while potentially improving the profile, economic competitiveness, and lived conditions within cities, are equally played out in the context of global, supranational, national and local structures, conditions, opportunities and constraints. While cities can improve their situations – the urban hierarchy is not fixed – they must construct their sovereignty in a global economy structured around competition and innovation. Understanding how consumption cultures (and the production of consumption) contribute to this competitiveness is the focus here.

## Consumption, governance and regeneration

The ways in which cities are governed as winners or losers and facilitators or respondents to this urban change have broadly been characterised through indicators relating to a generic shift of policy emphasis from local provision of welfare and services (collective consumption) to fostering local growth and economic development (individualised consumption). As Case study 7.1 shows, activities which were previously only associated with the private sector (such as risk-taking, inventiveness, place promotion and profit) have led to urban governance being variously described as new urban politics (Cox 1995), growth machines (Logan and Molotch 1998), urban regimes (Stone 1989; Handy 1994) or more popularly in academic literature as 'entrepreneurial governance' (Mollencopft 1983; Harvey 1989a; Leitner 1990; Hall and Hubbard 1998).

Case study 7.1 **A brief history of urban regeneration in the UK**

Throughout the eighteenth and nineteenth centuries, rapid urbanisation in the western world was marshalled by industrial paternalism, and burgeoning local authority power. Civic pride and Fordist industrial modes of production sought to support industrial growth through the provision of collective consumption. It is clear, then, that from the time of the industrial revolution manufacturing cities and their residents were clearly dominated by a small group of rich capitalists who organised the physical, social and political aspects of the city for the purpose of profit (LeGates and Stout 1996). For example, in the UK a plethora of government legislation, such as the Boards of Health and Local Government Acts, sought to bolster the activities of paternalistic industrialists and provide, for example, transport, utilities, health care, housing and education. The interests of bourgeois visions of urbanity were served as the city was developing around the needs of modern industrial growth throughout the first half of the 1900s.

More recently, from the mid-1960s to the present day, it is possible to identify four phases in the development of urban policy in the UK as a response to urban growth (Noon *et al.* 2000). In the 1960s, as part of the modernist 'systems' approach, programmes to tackle deprivation were very much based on the 'culture of poverty' thesis developed in America (see, for example, Banfield 1970). This attributed urban problems to the operations of families in small areas. Noon *et al.* (2000) show that urban renewal projects at this time represented an attempt to tackle 'social problems'. However, due to an overwhelming failure to make significant progress, Community Development Projects and Inner Area studies signalled a new direction that saw local problems being considered as part of wider structural economic change (and of course focused on attempts to come to terms with the problems caused by 1960s modernist planning).

This second phase ran up until the mid-1970s. However, the 1977 White Paper *Policy for Inner Cities* signalled another change in focus (and was followed by the change of government in 1979). This was a period characterised by an emphasis on the development of land and premises to bring about urban economic regeneration. This was also true of a third phase of urban policy, unfolded from the early 1980s until 1987/88. The Thatcher government saw local authorities as overly bureaucratised, and urban development corporations and task forces guided urban change instead (for a review, see Blackman 1995; Oatley 1998). Such policies were formed during a period of economic restructuring and increasing global interdependence, and a seeming rise in the power of footloose global capital. In the most successful (and aspiring) cities, local elites sought to make their cities attractive to the demands of global modes of capital accumulation – through speculative and often spectacular developments. Hence decision-making appeared to be disembedded from localities and based around globalised economic rationalities.

The fourth and most recent phase is essentially a mix and match of past urban regeneration philosophies. For example, Hambleton (1996) suggests that local authorities in the UK *are* now devoting more resources to 'economic' development – but while in some cases this is aimed at strengthening local attractiveness to global capital or wealth creation by local business elites, in others the focus is public service oriented, with redistribution policies seeking to economically empower the underclass. It is not that spectacular developments are not being undertaken – there was controversy during the early 1990s that money was often being siphoned from service provision to fund such projects – but that such projects are now being funded by European Union and National Lottery funds, not popularly seen as sources of direct taxation. Noon *et al.* suggest that programmes such as the Single Regeneration Budget now seek to undertake community regeneration (for example, employment and poverty-alleviation programmes).

One of the most striking features of this fourth phase has been the growing importance of urban policy that seeks to enhance the liveability of cities – and thus combining the goals of both economic growth and social justice. Local authorities seeking to enhance the cultural life of cities, through the redevelopment of the physical and cultural infrastructure of the city, often locate urban sociality and improved urban lifestyles for all social groups as key elements. This has been underpinned by the view that both spectacular culture-led regeneration and local communities' interests can be served by a more holistic policy approach. An attempt to promote such a joined-up approach to the development of UK cities was supported through the UK Government's White Paper entitled *An Urban Renaissance* (1999).

However, 'the jury is still out' as to whether many of the recommendations made in this report have had a substantial impact on urban life. Nevertheless, what is clear is that through the Department of Culture, Media and Sport, the UK Government at present pays considerable lip-service (and gives some financial support) to the idea that (in combination with other more traditional economic development strategies) cities can be regenerated via improvements to urban cultural life. For example, a recent document entitled *Culture at the Heart of*

continued

*Regeneration* (2004) described a raft of initiatives including iconic buildings, international and local festivals, and organic grassroots cultural programmes as being key elements in economically and culturally competitive cities. However, as Gibson and Stevenson (2004) argue, while rhetorical support of both speculator and grassroots developments is laudable, it is at the level of everyday policy decision-making and in the spending of scarce resources that the balance between social justice and a capitalist bourgeois vision of cities is generally sacrificed.

What is important to grasp from this brief review is that, whether discussing the city of the industrial revolution or the contemporary entrepreneurial city, theorising the relationship between political, economic, socio-spatial and cultural determinants is no easy task. As such, this chapter, in seeking to understand the shifting ground of the cultural economy of urban regeneration philosophies, must get to grips with producing an analysis which not only includes generalised theories of local and regional political, economic and cultural determinants, but also identifies how such broad urban agendas are articulated in the fabric of particular cities (and spaces/places within cities).

However, while there is undoubtedly a tendency towards repetitive outcomes (for example, in place-promotion campaigns and 'civic boosterism' that create repetitive and generic regeneration projects and hence 'blandscapes' in our cities; Short and Kim 1998), there are considerable variations in the entrepreneurial polices and projects which specific cities pursue. It is important not to deny the very different historical and contemporary trajectories of individual nations or cities.

The typology of the entrepreneurial city and the (re)imagining and (re)construction of cities in the contemporary urban hierarchy have been introduced as a new, more aggressive phase in the development of cities, with urban elites fostering projects designed to assert new forms of civic identity. Inherent in such an agenda is an attempt to legitimise the political projects that function to promote the economic and cultural interests of urban elites (Philo and Kearns 1993; Smyth 1994; Short 1996; Clarke, D. B. 1997; Pacione 1997). As such, the manufacturing of a new civic identity is undertaken and communicated through new inscriptions on the built fabric of the city itself. As we have seen, this has been particularly pursued through high-prestige or 'flagship' projects (Crilley 1993; Hubbard 1998; Short and Kim 1998), public art and furniture (Bianchini and Parkinson 1993; Goody 1994; Hall, T. 1995), hallmark events (Boyle and Hughes 1991; Ley and Olds 1988) and 'transitory topographies' such as sporting and temporary cultural events and festivals (Jarvis 1994), the construction of heritage centres (Hewison 1995) and through press discourses, posters and slogans (Brownhill 1994; Holcomb 1994).

Such practices themselves represent physical and symbolic attempts to produce suitable environments and infrastructure to sustain post-industrial/postFordist/postmodern economic activity and associated social and cultural identities and lifestyles.

However, a fundamental factor in the ability of cities to pursue such policies is the extent to which there is a successful replacement of a 'politics of legitimation' by a 'politics of coercion'. This suggests a movement away from local authority policies aimed at ensuring democratic debate and legitimation, to a situation where public–private partnerships become the central force dominating the political, economic, social and physical (re)construction of cities (Randall 1995). This has often led to local authorities part-funding or underwriting speculative developments; and, as we saw in Birmingham, this led to money being redirected from provision for housing, education and health to flagship projects and spaces/places within the city which may have social and cultural resonances for only wealthier residents and visitors. Speculative developments, moreover, rarely provide substantial 'trickle-down' benefits to all sections of the community (Leitner and Sheppard 1998).

It must be asserted, moreover, that not all cities have the capacity or ability to be successful in attracting global capital, developing financial and service economies, or providing the spaces/places to appeal to new residents, visitors and tourists. Not all cities have the institutional expertise to successfully instigate public–private partnerships with the potential to facilitate 'spectacular' developments or infrastructure improvements which produce the spaces/places and ambience associated with post-industrial/postmodern identities and lifestyles. Similarly, not all cities are sufficiently economically, spatially and/or socially diverse to innovate and significantly alter the physical and symbolic resonances of their past heritage; to render industrial and working-class employment and lifestyles marginal; or to socialise a significant proportion of the population to desire jobs, lifestyles and spaces/places associated with more successful cities.

## Consumption and urban regeneration in 'lesser' cities

If we consider the emerging regeneration strategies not of cities of global, national or regional importance, but of smaller urban areas it is possible to see stories of cities struggling to reinvent themselves in the context of de-industrialisation and the spectacular urban regeneration projects described in this chapter. An example would be the development of the city centre of a town in the English Midlands, Stoke-on-Trent (aka the Potteries), the centrepiece of which is the development of a Cultural Quarter. The ways in which Stoke-on-Trent is responding to current urban transformations and re-orientations offer interesting insights into particular social, spatial and economic conditions faced by small cities. By exploring the mesh of politics, policies and opinions which intersect around projects such as the Cultural Quarter, it is interesting to see how broad urban agendas are being woven into the fabric of a conurbation still deeply entrenched in traditional industrial lexicons.

Stoke-on-Trent has been woefully slow to respond to or develop in the light of global structural and institutional change. Economic and physical decay, unemployment, low wages, factory closure and a shortened working week characterise the remaining manufacturing industry. The poor image of Stoke-on-Trent is exacerbated by the comparable success of nearby neighbours Manchester and Birmingham and the progress of Nottingham and Sheffield. With a litany of decay and strong regional competition, the position of the Potteries remains bleak, with the conurbation apparently locked into a spiral of decline.

At the top of the urban cultural hierarchy, in cities such as New York, Paris, London and Tokyo, there is an intense competition for tourist money, firms, institutions and media events. In order for cities to compete, their ability to innovate and attract the widest variety of symbolic and economic activity, new spectacular experiences, events and spaces of consumption, is paramount. Lower-ranking cities, such as Manchester and Birmingham in the UK, for example, are less able to innovate, and thus impact predominantly on regional, national and perhaps supranational (European) flows of capital and culture rather than having a significant global influence. Cities such as Stoke-on-Trent are struggling to achieve even regional influence because of a seeming inability to compete or innovate in the symbolic economy. Weaker cities are forced to focus on local issues, often pursuing unclear development potential or internal reorganisation (Herrschel 1998). While this may have some success in making particular areas of the city socially, culturally or economically distinct, such practices have little impact (or success in terms of specific market segments) upon attracting or influencing flows of capital or culture outside the city, region or country.

However, in spite of this apparent political, economic and cultural inertia, September 1999 saw the culmination of a nine-year process of consultation, with successful funding bids for around £20 million from the National Lottery and European Regional Development Fund, and Stoke-on-Trent's Cultural Quarter project was under way. The formulation of the concept of the Cultural Quarter can be seen to emerge from the 1990 report *A Cultural Strategy for Stoke-on-Trent*, which concluded that for a city of over 250,000 people and a catchment of a further 925,000, there had been under-investment and a lack of 'arts and media'-oriented infrastructure which many similar towns and cities take for granted.

Stoke-on-Trent was failing to satisfy the needs of its population, hence the report recommended the need for a distinctive cultural quarter in the city.

The overall vision of the Cultural Quarter is based on the redevelopment of the two disused theatres, support for existing facilities (including a cinema – which has now shut down – a youth theatre, the municipal museum and art gallery, and the city library), and the augmentation of the area by enhanced retail activities. July 1999 also saw the beginning of the city-centre public realm project, a three-year, £1.6m initiative funded principally by the European Regional Development Fund giving Hanley 'a make-over' (Holcomb 1994). However, despite the opportunity (and optimism) that the public realm project had to offer, the results of the project are less than spectacular. Unlike the redevelopment of, for example, Birmingham's Centenary Square, Leeds's Millennium Square or post-bombing Manchester in the UK, the public and civic spaces created by the public realm project are far from impressive (see Figure 7.6).

Moreover, the continued dominance of the Potteries' industrial past is startlingly evident in an almost overwhelming reluctance to positively represent post-industrial economic development. For example, there has been very little progressive championing of the Cultural Quarter, the potentialities of the post-industrial economy, or the activities of cultural intermediaries in either official or media discourses. Representations of the city centre in the local press depict images of

**Figure 7.6 _Stoke-on-Trent's Cultural Quarter. (Photo: Mark Jayne)_**

crime, decay, empty shops, homelessness, rats and drunkenness. There appears little willingness to positively represent or support identities, lifestyles and consumption practices of, for example, lesbian and gay, ethnic and other social groups who have been so intimately related to the revitalisation of other cities. Quilley (1999), for instance, identifies how there has been a tentative embracing of Manchester's Gay Village by the city council as it promotes the city abroad. In Stoke-on-Trent, *MoveMag*, a local free listings magazine, is the only local publication to actively promote post-industrial activity and mix-and-match lifestyles and consumption practices, and to positively represent the activities of cultural intermediaries such as local bands and DJs, and local cultural producers such as artists and craftspeople. It is suggested here that the continued stubbornness of local vernacular associations ensures that there is currently a no-go area of representation in which the promotion of identities and lifestyles associated with the post-industrial economy is considered pretentious, yuppyish or a threat to political, economic or social continuity (see Wynne and O'Connor 1998, and later in this chapter).

As such, while the presence of arts, entertainment, food, music and fashion-oriented development may suggest processes of gentrification, the attraction of employers and the promotion of lifestyles associated with the managerial, professional and service economy is being less vigorously pursued by a local authority which has been woefully slow to acknowledge the importance of the interface between culture and economic development. Accordingly, the Cultural Quarter seems a hopeful rather than tangible and practical attempt to stimulate post-industrial activity.

In contrast, Manchester city centre's success in its property-led regeneration has stimulated broader leisure, tourism and business investment. Quilley (1999) shows that this emphasis on 'bricks and mortar' renewal and re-imaging the city as distinctive and individual was undertaken by concentrating on flagship projects in the city centre, which would then be perceived as successful. In a strong economy of inter-urban competition, Manchester sought a 'market niche' that, while being focused around the conventional financial and professional services, culture, media and leisure industries, and higher education, exploited the city's image as a 'cool' place. In addition, Mellor (1997) shows that promotion of a '24-hour city', stimulated through the relaxation of licensing and planning policies, was undertaken in Manchester in conjunction with promotional, financial and political support of the already thriving culture industries.

In other cities, the creative and cultural industries have now become central to attempts to develop post-industrial city-centre cores. This sector is considered to be important not only because of its own burgeoning economic wealth but also because it creates a buzz within cities. The buzz is about creativity, entrepre-neurialism and innovation. Such conditions are attractive not only to middle-class gentrifiers (or working-class bohemians; Milestone 1996), but also to capital and investment. There are now many examples of cultural quarters throughout the world; in the UK most cities now have these initiatives as part of their repertoire of urban regeneration schemes. However, what makes these quarters successful is the targeted aim to stimulate the production and consumption of post-industrial economic and cultural activity.

For example, Sheffield's Cultural Industries Quarter (CIQ) is a small, well-defined area to the south-west of the city, within five minutes' walking distance from the city centre. A number of arts organisations were already based in the area in 1984 when Sheffield City Council began a series of initiatives to attract cultural enterprises. The area now houses many independent small to medium sized enterprises as well as larger organisations like Sheffield Science Park and Sheffield Hallam University. Figures based on an independent audit in June 1995 show that there are 109 cultural businesses and organisations located in CIQ, and it has created 770 full- and part-time jobs and provides 1265 training places per year. The combined turnover of the businesses and organisations was £20m in 1996/97. There has been no such support for creative industries development in Stoke-on-Trent's Cultural Quarter.

In the highly symbolic post-industrial economy, where a city's image and profile are central to its ability to attract flows of capital and culture, as we have seen, place promotion has become increasingly important and aggressive. In terms of Stoke-on-Trent's city centre, the low-key campaign 'Stoke-on-Trent Deserves its Own West End' (promoting the Cultural Quarter) and the promotional video *Stoke-on-Trent City Centre: A Better, Safer Place to Be* are the only projects. More recently, in the light of severe flooding in the UK during 2001, the City Centre Management's literature adopted a tagline 'at 700 ft above sea level the City of Stoke-on-Trent is probably highest in UK'. While this indeed may (probably) be a relevant selling point, it will (probably) not have the same success as 'I♥NY' or 'Glasgow's Miles Better' (see Ashworth and Voogod 1994; Gold 1994; Revill 1994; Ward 1994, 1998; Brown 1995). In sum, the modest and low-key place promotion of the city centre highlights a lack of promotion, branding and marketing expertise which is not being addressed.

It is perhaps not surprising, then, that while Stoke-on-Trent is the eleventh largest conurbation in the UK, in terms of investment the city centre is currently ranked only thirty-third (City Centre Management 1998b: 6). At present, Stoke-on-Trent is ranked only the fifty-fourth most popular city for business relocation in the UK (Bradford and Bingley Relocation Service, in *The Sentinel* 1999: 28). Moreover, in Stoke's city centre, the most prominent spectacular events are Peacekeeper (a celebration of all things military with marching bands and simulated exercises), the Lord Mayor's Parade, the Parade of Bands and the Potteries Heavy Horse Parade. While there has recently been the more participatory and potentially more 'carnivalesque' Afro-Caribbean Carnival and, in 1998, the one and only running of the 'Loud and Proud' Lesbian and Gay Festival, these do have the same profile. These pageants seem to pay homage to industrial and nationalistic working-class discourses, rather than a more cosmopolitan urban culture and the '[E]uropeanisation of street and café culture' seen extensively in most other cities (Arts Council of England 1995: 5).

Similarly, while Miles (1997) identifies that public art is used to embody discourses of new post-industrial urbanism, again Stoke-on-Trent has been slow to adopt such agendas. In terms of public art, the £80,000 project the Pleasure Gardens of the Utilities (two long ceramic-covered benches with a blue and white floral design, mature silver birch trees and, during the first three months, a video screen showing portraits of local people) has been the most significant. In response to this deeply unpopular project, which was popularly considered as 'a lot of money for a few benches and trees', *The Advertiser* (1999a) ran a poll to assess public opinion – the results were No: 254, Yes: 35 to the suggestions that any further such projects should be undertaken. However, a £1.8m Lottery bid by the city council to produce a string of art installations around the city failed and the public art policy is thus in disarray, and at the time of writing no one person or council department was responsible for public art policy. The city centre has, however, been the focus of several projects such as 'Sounds of Industry to Pervade the City', where shoppers were serenaded with sounds from potbanks, pits, steelworks and other industries, and a siren at lunchtime (*The Advertiser* 1999b); and the 'Bringing a Touch of Colour' project, where concrete slabs were painted bright colours, turf was laid and planting undertaken on a traffic island, and a pedestrian bridge was decorated with terracotta-coloured banners (*The Advertiser* 1999c). Unfortunately, within a few months this colourful oasis had declined, the paint had faded, the turf became a muddy walkway, and the banners had blown down.

The development of this urban village thus appears flawed in two ways: first, it could be argued that the development of the Cultural Quarter has been initiated to stimulate the growth of a critical infrastructure of cultural producers and consumers and that this can be regarded as a 'shallow' attempt to stimulate post-industrial activity. While the presence of arts, entertainment, food, music and fashion-oriented development may suggest processes of gentrification, the attraction of employers and the promotion of lifestyles associated with the managerial, professional and service economy is being less vigorously pursued. Secondly, it would appear that there is little will (or perhaps ability) to generate a distinct social or spatial signature not reliant on 'traditional' industrial lexicons.

The Cultural Quarter has not emulated London's Covent Garden, or Brighton's Lanes, Nottingham's Lace Market, Manchester's Northern Quarter or Liverpool's Rope Walks, and developed a thriving sector of 'trendy' designer shops and a café culture. Indeed, the titles of Bar La De Dah (café bar) and Toffs (sandwich bar) show how there is some ambivalence towards wholly promoting 'posh' middle-class lifestyles and iconography in the midst of a working-class city. While there are some 'lifestyle' shops and trendy clothes shops, the Cultural Quarter area is still dominated by charity shops, cafés, pubs and shops that include a pawnbroker's, a bookie's, a butcher's and a militaria shop. The consumption opportunities in the Cultural Quarter are thus best described as partially a locus of Stoke's own blend of postmodern lifestyles, identities and forms of sociability, but working-class production/consumption cultures continue to dominate the city centre. At present, unless 'low-key' becomes the next 'spectacular', then there is little of the conurbation's social and physical landscape to attract or contribute to wider flows of capital, culture and people (see Figure 7.7). However, what remains clear from statistics on unemployment, dereliction and lack of investment is that Stoke is a city that is continuing to underperform.

Thus, while entrepreneurial urban practices attempt to produce attractive and distinctive locations in the face of intense inter-urban competition, what must be stressed is the extent to which the characteristics of particular cities (or

**Figure 7.7  A typical shopping street in Stoke-on-Trent. (Photo: Mark Jayne)**

places/spaces within cities) dictate their degree of (potential) success in the post-industrial economy. Many cities (and areas within cities) are less concerned with economic growth than economic stability, or with halting a spiral of decline by producing more 'mundane' regeneration projects. This often involves not 'spectacular developments' but rather cosmetic renovation of declining city centres; attempting to produce a competitive advantage in localised economic, cultural or spatial terms; or simply by 'greening' initiatives, community projects or retraining in order to improve the job prospects of the (post)industrial workforce. Such diversity in the entrepreneurial orientations of urban governance highlights how issues of scale, scope and efficacy must be central to any conceptualisation and evaluation of urban renewal projects (Hall and Hubbard 1998). This is not only an important goal in itself, but also highlights that we must be aware of how local cultures mediate, contest and negotiate their cities' trajectories and how different groups, with differing mindsets and agendas, view cities.

Hence, it is important to understand that the willingness and ability to pursue sustainable entrepreneurial activity are socially embedded in the complex world of political, economic, social and physical opportunity (Loftman and Nevin 1998). The way in which political and public discourses collide and conflict leads to negotiation over issues of economic competitiveness and the social and cultural orientations of collective consumption. Many examples of over-adventurous and unsustainable entrepreneurial policies have led to taxpayers picking up the cost of speculation and uncertainty (Harvey 1989b; Beck 1992). 'Best-practice' blueprints are regularly replicated in 'me-tooist' strategies, where any advantage is ephemeral. While there is a ubiquitous 'narrative of enterprise', necessarily designed to promote policies of local growth and attracting inward investment and jobs, Jessop (1997) identifies different *levels* of entrepreneurial activity. These surround the spatial and social orientation of particular urban economies, and their ability to attract economic and symbolic capital and to provide local/regional/national/global cityscapes with geographies of consumption which (re)define the urban hierarchy at specific levels.

The often glaring disparity of investment in many cities has led to community, employment and small-business-oriented projects, and less spectacular entrepreneurial activity designed to alleviate deprivation and poverty and to enhance the liveability of cities. These economic and business-oriented initiatives include reskilling the workforce, enhancing local and regional gateways, and developing new sources of supply to enhance a city's economic base (Painter 1997). Public–private partnerships have also been activated to pursue economic development and community (or collective consumption) oriented projects (Jacobs 1992). The funding of ethno-religious initiatives (Ball and Beckford 1997), crime management projects (Hughes 1997; Lea 1997), health and poverty programmes (Moore 2000;

Sheaff 1997), community centres, the physical enhancing of problem estates, greening and infrastructural improvement to housing (Radcliffe 1997), street furniture and CCTV, the targeting of employment black spots and problem estates (Foster 1997; Mooney 1997), and projects to improve participation in local politics have all been initiated in this context (Beazley *et al.* 1997; Geddes 1997).

It is clear, then, that there is no 'quick recipe' for successful entrepreneurial policies and practices. Similarly, the potential for successful mobility, with those dominant cities aggressively exploiting past success, is limited. Even middle-ranking cities (such as Dublin and Barcelona) have few opportunities for economic expansion in the face of stubborn centre–periphery relationships (Jessop 1998). Thus, cities develop strategic approaches that take into account strengths and weaknesses. Cities such as Cheltenham, Swindon and Bristol in the UK are not growth-oriented, concentrating instead on attracting market segments of high-tech industries and providing a residential environment that satisfies the needs of professional and managerial workers (Hall and Hubbard 1998). Similarly, the variety of innovative techniques for generating liveable and creative projects is site- and opportunity-specific (Landry 1995). Moreover, the political climate is volatile and policies are changeable. This can be exemplified by Birmingham's development of several flagship projects to try to compete more effectively as a middle-ranking European city.

It has been argued that in the contemporary urban hierarchy, cities strive with varying degrees of success to strengthen their identities as consumption centres, and that quantitative and qualitative differences between them thus come to the fore. In order to understand this uneven development, the chapter has argued that there is a need to focus on such spatial outcomes as complex mediations. For example, Castells and Hall (1994) and Hall (1998) note that the overwhelming focus of capital accumulation has moved from mines and foundries, to an information- and knowledge-based economy, and from an industrial to a post-industrial (Fordist to postFordist) urban hierarchy. However, this move is not predestined. They further suggest that territorially defined grassroots organisations and local services can preserve their identities and have significant control of the discursive construction of localities. Nevertheless, those cities and regions that do not find a specific role and place in the new information, networked and symbolic economy will also struggle to be competitive.

Taylor *et al.* (1996) suggest that in many older, previously industrialised cities recent changes have indeed brought about a reworking and re-representation of the focus of local identity and differentiation. Cities can maintain, improve or lose their position in the urban hierarchy, and this is dependent on their ability to re-imagine their political, economic and social and cultural identities. This can include developing new sectors of the economy and new places/spaces of consumption, or by promoting post-industrial/postmodern identities and lifestyles (albeit in a modified fashion with hybrid customs and practices relating to the modern period). Others, on the other hand, can be complicit in failing to compete at all.

Nevertheless, Peter Hall (2000: 640) suggests that as nations and cities have passed from a manufacturing economy to an informational economy and from an informational economy to a cultural economy, '[c]ulture is now seen as a magic substitute for all the lost factories and warehouses and as a device that will create new urban image, making the city more attractive to mobile capital'. Hall notes that 'historically the most talented and creative milieux are often uncomfortable, unstable cities, or places of conflict, and [that] typically creative people are from localities somewhere removed from centres of power and influence – but not so distant to be cut off' (2000: 643). However, while this may indeed be true, cities can improve their position, and Hall is right to show that cities have historically risen and fallen – it is generally the rule that it has been the cities with a diverse complex of economic activity, social and cultural diversity, and associated spaces/places of consumption that have been consistently sustainable.

## Consumption and the urban hierarchy

A key issue here in understanding the ability of cities to compete in the contemporary urban hierarchy is the question of citizenship, and the way that the consumer citizen is a central figure in debate about the success of world cities and the practices of cosmopolitanism. This focus on consumer citizen is particularly important in terms of the production of space (Bell and Binnie 2000). For example, it is clear that consumption is central to the ways in which citizenship is defined, and the construction of our identities, and the management and disciplining of self, occur through choices we make as consumers. As was shown in Chapter 6, 'freedom' and 'power' are increasingly articulated through the market (Bell and Binnie 2000).

However, what, and how, consumption takes place and the associated power relations in specific contexts need further elaboration. As established earlier, while it is clear that the changing role and success of the state – and specifically the local state – rests on making cities more desirable places, it ultimately depends on local conditions, opportunities and constraints. For example, Richard Florida (2002) describes the presence of a 'creative class' as being key to the ability of cities to remain competitive, and highlights the important role of the state in fostering the spaces of/for consumption for such a creative class to act entrepreneurially. This process, as we have seen in the previous chapter, marks those who are desirable and those who are undesirable consumers. In Florida's terms this group is a mishmash of different, generally middle-class subgroups made up of artists, lesbian and gay people, transnational venture capitalists and cosmopolitan tourists. Florida argues that it is such people who can actualise the 'city of the future' – framed as vibrant, cosmopolitan, entertaining and 'happening' (Mellor 1997). But does the presence of such groups necessarily bring success for those urban areas that possess such groups in significant numbers?

It must be argued that competitive urban conditions are very much dependent on factors such as a strong regional economy with a professional post-industrial business core; domination over the counter-magnet of the suburbs; and the stimulation or maintenance of a 'critical infrastructure' of citizens. Importantly, the mix of people who inhabit a place is a major determinant of success. What must be stressed is that it is not just bricks and mortar regeneration projects but social relations and forms of sociability that produce a significant competitive advantage or specialism in the post-industrial economy.

Massey's (1993) conceptualisation of power geometry and space is an important argument here. Massey shows how individuals, groups, cities and regions are placed in very distinct ways in relation to the flows and interconnections of political, economic, social and cultural process of globalisation. Some initiate flows while others are on the receiving end of them. So, those who are 'doing' the making and the communication are in a position of centrality. Thus, in terms of seeking to understand the transformations (or not) of cities, consciousness of links with the wider world (which integrates in a positive way local, regional, national and supranational political, economic and cultural influence) is neces-sary. In presenting a more 'progressive sense of place', Massey highlights how theorising the transformations of cities is about successfully conceptualising a global construction of place. What is important here, then, is that while cities are presented as nexus points for flows of capital and culture, Massey depicts them as dynamic, social constructions – the result of historical processes where different political, economic and cultural practices have sculpted the character of the city.

Such an understanding suggests that the ability of cities to compete at particular nodes in the post-industrial urban hierarchy is in part delineated by geographies of consumption (and production) that operate at particular spatial scales. For example, global/supranational/national/regional cities are such because of the physical and symbolic infrastructure, and the political, economic, social and cultural facilities and structures that they developed and sustain. In essence, global cities have the highest concentration of financial and business institutions; political and economic centrality; and the right mix of lifestyles and social relations – which, importantly, *represent* those cities as socially and culturally successful, innovative and attractive. The social and spatial re-organisation of capital, which has resulted in new functions for financial markets, has created this re-ordering of the significance and influence of cities across the world. However, this has also ensured that while some cities can successfully compete in this post-industrial economy, others are disadvantaged. In essence, some intra-city spaces/places and regions monopolise control functions, while others are subordinate (Massey 1993; Mercer 2000).

While this is perhaps understandable, as academics seek to promote breakthroughs in critical discourse, it has been cities near the top of the urban hierarchy – global cities such as Los Angeles, London, Singapore and Sydney, or cities in the western industrialised world with national or regional significance, such as Barcelona, Indianapolis, Manchester and Birmingham – which have mainly been the focus of theorists. However, as Thrift (2000) states, cities are not homogenised entities – not all cities are the same; not all cities (or spaces/places within cities) compete at the same social or spatial level; one story of urbanity cannot tell all.

In a similar vein, research has also tended to be overly deterministic in its theorisation of factors such as the social construction of the social relations, identities, lifestyles and forms of sociability that contribute to the nature of everyday life in our cities. In sum, the ways in which space and place make a difference to their discursive construction has been under-theorised (Savage and Warde 1993; O'Connor and Wynne 1996). When studies consider archetypal identities which have been associated with urban change (whether they be new petite bourgeoisie, industrial working class, bohemian artists, immigrants, lesbian and gay, youth and ethnic social groups – or any other consumption identity couplet which could be mentioned), it must be understood that although there are groups of consumers who share similar lifestyles across national boundaries and continents, there are also important differences between them (Glennie and Thrift 1992).

Wynne and O'Connor (1998) graphically show this through their study of a group of people who have recently moved to Manchester city centre (UK), in order to discuss consumer cultures and the postmodern city. By using ethnographic techniques and questionnaires they examined cultural consumption, lifestyle and forms of sociability – such as theatre-going, gallery and film attendance, and musical and political preferences. They also use qualitative analysis of features associated with changing identities under conditions of late or postmodernity (these include self-monitoring or reflexivity, changing patterns of sociability, and the aestheticisation of everyday life). While debates on globalisation have focused on information and commodity flows and how these impact on sense of place, there has been less concentration on how city cultures respond to this. Wynne and O'Connor thus seek to understand how a cultural investment for urbanites, 'living on the edge' (Shields 1992b) and moving to the city centre, is both a generalisable 'distinction' strategy and is constructed differently in the context of particular cultural contexts (Zukin 1982, 1995).

Wynne and O'Connor (1998) found that cultural intermediaries were an increasingly diverse group, and that Manchester's vibrancy was very much based around its music scene. Contra to Zukin, the centrality of identity worked in different ways to create lifestyles and 'habitus'. This particularly related to what it is like to be a 'northerner', or to live in a 'northern town' (Shields 1991). Wynne and O'Connor (1998) sought to identify markers of place-based identity, and how those meanings related to wider social practices and patterns of social life rather than an accretion of cultural capital. The traditional pattern of a rising social grouping aspiring to the cultural capital of a higher one, whether through imitation or out-flanking, is no longer a viable model for Wynne and O'Connor (1998). In their case study, cultural capital does not ensure a primary means of ensuring distinction, nor could it be regarded as avant-garde or a taste marker in the field of fashion and lifestyle.

Moreover, the research found enthusiasm for middle-brow levels of cultural capital, and a distrust of 'pretentiousness', despite knowledge of higher cultural goods. There was a tangible division between high culture and popular culture, which provided structural markers. Regarding the provincial–capital and north–south divisions, Wynne and O'Connor suggest that the reluctance to over-play cultural competition and/or resentments points towards a refusal of a cultural model associated with London and resistance to pretentiousness that marks northern culture (see also Taylor *et al.* 1996). The transformation of city cultures is complex and demands contexts (and historical accounts). The debate on the relationship between consumption and urban regeneration needs to recognise that place as well as space 'makes a difference' (Wynne and O'Connor 1998).

## Concluding remarks

It has been shown in this chapter that cities throughout the western world are seeking to improve their economic competitiveness by initiating cultural- and consumption-led urban regeneration programmes. If cities are to compete in an urban hierarchy dominated by post-industrial economic activity then they must attract flows of capital and people – investment, jobs and tourists. A vital factor in their ability to do this is the range of consumption opportunities that they offer. What must be remembered, however, is that urban regeneration projects are undertaken in the context of very specific social relations, and while we can generalise about broad urban agendas, there is great localised diversity that constitutes the seemingly standardised constructions of consumption and production cultures.

Nevertheless, consumption-led urban regeneration programmes can be characterised by the spread of bourgeois globalised production and consumption cultures, and the diverse cultures that constitute our cities are generally marginalised in such projects. Moreover, the building of flagship cultural projects and iconic buildings often leads to city authorities redirecting funds away from welfare programmes aimed at ensuring social justice. Not all cities, however, are able to overcome the economic and social problems associated with de-industrialisation. Many cities, in attempting to initiate consumption-led urban regeneration programmes, simply highlight both their own lack of progress and the relative advances, or continued economic and cultural success, of other cities.

---

**Learning outcomes**

- **To have an understanding of the relationship between consumption and urban regeneration**
- **To be able to describe the ways, and give examples of how, consumption is utilised in urban regeneration projects**
- **To offer insights into why consumption cultures are supported by local governments**
- **To be able to critically appraise the ability of cities lower down the urban hierarchy to utilise consumption-led regeneration in order to improve their competitiveness**

---

## Further reading

Richard Florida (2002) *The Rise of the Creative Class*, New York: Basic Books. A much hyped book and well received in the policy arena and in the media, but less well by academics. The book provides empirical evidence for the importance of a creative class to urban life, looking at issues such as work, life and leisure and community.

Tim Hall and Phil Hubbard (eds) (1998) *The Entrepreneurial City: Geographies of Politics, Regimes and Representation*, London: John Wiley. A very useful review of strategies labelled as entrepreneurial local government policy and projects. From flagship projects to crime and health, contributors discuss the merits of and effects of local authorities engaging in 'risky business' and its effects on service provision.

Malcolm Miles (1997) *Art, Space and the City*, London: Routledge. An excellent and comprehensive critical review of the way art and design are integral to urban life. The book reviews the role of art both in regeneration and as a social process of criticism and engagement, with a wealth of examples from around the world.

David Bell and Mark Jayne (eds) (2004) *City of Quarters: Urban Villages in the Contemporary City*, Aldershot: Ashgate. This edited collection is an integrated primer that provides theoretical and empirical discussion of urban quarters throughout the world. Chapters are focused on topics such as production and consumption, urban regeneration and identity, lifestyle and forms of sociability.

Graeme Evans (2001) *Cultural Planning: An Urban Renaissance?*, London: Routledge. A detailed and wide-ranging review of the way in which cultural life, planning and policy is a key part of urban life and has become central to regeneration initiatives around the world.

# 8 Conclusion

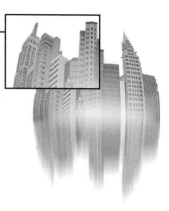

**Learning objectives**

- To briefly outline and summarise the key themes covered in the book
- To revisit important debates and case studies and draw links across the chapters
- To highlight a future research agenda for studies of cities and consumption

This book has investigated the mutual and dynamic relation between urban development and consumption, and the ways in which cities are moulded by consumption and consumption has been moulded by cities. It has been argued that consumption stands at the intersection of different spheres of everyday life – between the public and the private, the political and the personal, the individual and the social. Consumption has been shown to be a means and motor of social change; an active ingredient in the construction of space and place; and central to our identities. It has been argued that consumption has multiple political, economic, social and cultural roles, and that it is in the morphology of cities that its expression is most explicit.

This chapter will briefly outline and summarise the key themes and topics covered in this book and revisit important debates and case studies in order to draw links across the chapters. The chapter will conclude by signposting a future research agenda for studies of cities and consumption.

## Key themes and topics

From a relatively marginal place in social sciences, consumption has become central to our theoretical and empirical understanding of the world. From being associated with waste, negativity and 'using things up', the consumer is now an active citizen, and consumption is a principal way to participate in civil society. However, our understanding of consumption and its role in urban change has evolved over time. Many of the different theoretical and methodological approaches that seek to explain the relationship between consumption and urban change have been described throughout the chapters of this book.

Chapter 1 began with an introduction to the various ways in which consumption has been defined. This review highlighted that everyday consumption underpinned the development of our consumer society. The relationship between consumption and urban change was then discussed in terms of the work of Marx, Veblen, Simmel and the Frankfurt School, and concepts such as collective consumption. This was followed by an introduction to the ideas of more recent theorists, such as Baudrillard, Bourdieu, Featherstone, and Lash and Urry, who sought to describe the relationship between postmodern urban life and consumption.

The chapter then investigated what it is we think about when we think about the city. The relationship between urban change and concepts such as postFordism, the postmodern and post-industrial capital accumulation were explained. Subsequent sections introduced Stephen Miles's (1998a) concept of the 'consuming paradox', which shows that consumption allows us to have a sense that we are free to construct our own identities, but that consumption also structures and constrains our lives. This was followed by an assertion that consumption can be studied both as a general topic that helps us to describe global practices and processes and also in more specific ways to explain local differences. However, in order to fully understand the complex interaction of political, economic, social, cultural and spatial practices and processes, consumption cannot be studied in isolation. It was argued that the 'circuit of culture' (Du Gay 1997) provides a useful framework to show that consumption is inextricably bound up with production, regulation, representation and identity. The chapter then concluded with a review of the various and often competing theoretical traditions and methodologies that have attempted to explain urban change.

Chapter 2 focused on the relationship between consumption and the development of the modern city. It was in the modern city that consumer culture and the core political, economic, social and cultural institutions, organisations, infrastructure practices and processes of modern life were inextricably brought together to a degree and extent never seen before. This complementary strengthening of political institutions and organisations, and economic growth, and the parallel development of cultural- and consumption-oriented activities dramatically impact on the development of the modern cityscape. It was argued that the relationship between consumption, modernity and the development of the modern city was founded on rational organisation, planning, scientific know-how and technological advances.

Equally important is the understanding that it is in the modern city that consumption becomes bound up with the very idea of the modern world. Consumer culture is therefore not simply founded on the industrial and intellectual success of modern thought – but rather 'the consumer' and the experience of consumerism are integral to making the modern city. The work of Marx, Veblen and Simmel was discussed in detail in order to show that the idea of consumption was not merely as an add-on to studies of production, but as a marker of social prestige and central to the development of urban life, deserved theoretical attention. The 'mass cultural critique', or the production of consumption argument advocated by the Frankfurt School, was debated. The Frankfurt School suggested that the rise of mass production in the twentieth century led to the commodification of culture. Consumption thus served the interests of manufacturers seeking greater profits, and in this process citizens became passive victims of advertisers. The shift is often associated with the decline in the public sphere and growing privatisation.

Sections in this chapter also described how consumption was central to bourgeois attempts to control the unruly working classes and impose social and spatial order on the city. Le Corbusier's modernist vision of the 'city as machine', as well as theories surrounding collective consumption and the idea that the city could be described as a model, were outlined in detail. Construction of modern identities, lifestyle and forms of sociability were discussed via archetypal urban spaces such as department stores and arcades and elaborated through the figure of the flâneur. The urban conditions associated with late modernity and problems surrounding economic depression, the increased social and spatial polarisation of the city and the growth of ghettos and the suburbs were then depicted.

Chapter 3 showed that over the past twenty years the nature of everyday urban life has been profoundly affected by the global reconstruction of economic, political, social and cultural processes. Related to this profound change was the decline of the heavy and manufacturing industries that dominated the modern city and an increase in the importance of post-industrial service industries such as financial services, banking, advertising, marketing, public relations and the retail sector. This was coupled with social and demographic forces that saw the simultaneous increase in mass unemployment, and the rise of a new middle class. These changes were argued to have been bound up with a movement to late or advanced capitalism – from modern to postmodern times.

One of the central tenets of this conceptualisation of a 'new' global spatial and symbolic urban economy and hierarchy is that the city, which historically was politically, economically, socially and spatially organised around production, is now said to be underpinned by consumption. The significance of culture was linked to the rise of a symbolic economy concerned with making and distributing images.

In the postmodern city, projection of image lies at the heart of the attractiveness and style in the city. In contrast to modern cities, where function shaped appearance, and products and buildings were mass-produced and generally standardised, in the postmodern city style, design and appearance rule.

This genealogy of the people and places that characterise the postmodern city provides a valuable generalisable template. It was argued that such a template allows us to identify how broad urban agendas are articulated in the fabric of particular cities, as well as providing a framework for comparative research. Nevertheless, what depictions of the modern and the postmodern city show is that no one story 'fits all'. It was suggested that our understanding of the relationship between cities and consumption must therefore be posited in both general and specific ways and must be informed by both theoretical and empirical research.

The chapter focused on the work of Bourdieu (1984), who showed that consumption of symbolic goods signified prestige, status and social standing. He argued that consumption was about a process of identification and differentiation, and that consumption was part of the way we articulate a sense of identity. Our identity is made up through the consumption of goods and services, and their display constitutes our expression of taste. This argument was taken up by postmodern theorists who argued that culture had become increasingly fragmented, and that the symbolic role of consumption had become increasingly important. Baudrillard, for example, suggests that we become what we buy. He argued that it is signs and signifying practices that are consumed, but that signs have no fixed referent – as such, any object can in principle take on any meaning. In Baudrillard's terms, then, we are left with a society of pastiche – underpinned by a play on signs with no reference beyond the commodity.

The physical, political, economic, social, spatial and cultural changes associated with the postmodern city were then described. Concepts such as Fordism and post-Fordism were discussed and archetypal postmodern urban spaces such as shopping malls, urban villages and gated communities were unpacked. The role of the state in supporting particular urban consumption cultures was discussed. The chapter concluded by seeking to disrupt archetypal descriptions of the postmodern city by highlighting how local political and economic conditions, social relations and cultures mediate the spread of globalised urban conditions in specific ways.

Chapter 4 acted as an important antidote to depictions of spectacular urbanism. The focus was on 'ordinary' or 'mundane' consumption, and 'everyday' urban spaces, activities and social relations. This included discussion of inconspicuous consumption spaces such as car boot sales, charity shops, retro/second-hand clothes shops, markets, supermarkets and the home. The ways in which people engage with the spectacular urban landscapes in 'ordinary' and mundane ways was grounded in a review of the work of Michel de Certeau and Henri Lefebvre. The ways in which individual agency relates to consumption practices, and weaknesses in the more structurally biased urban studies of archetypal urban spaces/places and identities, lifestyles and forms of sociability, were highlighted.

It was argued that everyday life can be seen as productive consumption – that consumption is not an end process, but the beginning of another form of production. Sections investigated a number of everyday consumption practices and spaces – for example, the social and cultural role of the home and the kitchen table, eating out, and supermarkets. The chapter also looked at the second-hand city, with a focus on car boot sales and charity shops. A discussion of the consumption practices and everyday lives of older people in the city concluded the chapter.

Chapter 5 showed that while studies of urban consumption have predominantly focused on middle-class consumption cultures there has also been a growing focus on other urban identities. Sections looked at the experience of being poor in a consumer society and how extremes of 'haves' and 'have-nots' are written on to urban landscapes. Gendered consumption, ethnicity, sexuality and subcultural style were discussed in detail. The chapter showed that consumption and the opportunity to consume are key indicators of power structures and are a key part of how social hierarchies, relations and everyday practices and processes are constructed. Moreover, it was argued that consumption can be understood as playing a key role in codifying power as social space in the city. In 'contested' spaces in the city, identity is constructed and regulated, and the social construction of difference, negotiation of identities and power struggles take place. It is in these terms that codifying and interpretation of the city through studies of consumption involve knowledge of actual usage, power relations and discourses in everyday life.

Sections focused on constructions of working-class carnivalesque consumption practices; gendered consumption practices in department stores; and ethnicity and consumption through food, fashion and music. The chapter also considered the relationship between consumption and constructions of sexuality and subcultural style. The chapter highlighted that while these issues can be theorised at a general level they can only be more fully understood and established with reference to empirical groups and contexts. It was argued that urban studies of consumption must ground accounts of emergent lifestyle groups in terms of specific places and spaces.

Chapter 6 argued that the city, and spaces and places within it, not only are sites of consumption but are also themselves consumed. Representations of the city in the virtual world of consumer society and the ways in which the city is represented in place promotion, planning and other official discourses were discussed. The ways we consume spaces and places – through sight, smell, sound and touch – all of which have an effect on how we interpret and experience the city, were also depicted. These topics were framed through the work of Henri Lefebvre and Edward Soja, who highlighted how the imaginary city and the ways we consume urban spaces and places not only affect the way we experience, but inform the material development of our cities. An important argument throughout this chapter was to show that the city is produced and consumed in multiple ways. There is no one single representation, interpretation or consumptive experience that can be said to know, describe or encapsulate a city, spaces or places within it, or aspect of urban life. Our interpretation and experience of consuming the city, whether it is real or imagined, is very much dependent on social and hence power relations that surround constructions of class, gender, ethnicity, sexuality, age, nationality, and so on.

Chapter 7 elaborated the view that in a global urban hierarchy characterised by intense competition, cities are promoted and sold not simply as centres of economic growth but as culturally rich places in which to live and work, where the quality and quantity of consumption opportunities are crucial elements. The chapter critiqued the seemingly ubiquitous presence of consumption-led urban regeneration initiatives and the implications for cities throughout the urban hierarchy. The proliferation of spectacular urbanism, with case studies including Birmingham, Singapore and Chicago, was critically debated. The role of government in supporting urban regeneration and consumption-led regeneration strategies and the strategies of lesser cities were discussed.

## Cites and consumption: towards a new research agenda

This book has shown that today consumption is not seen as corrupting, nor are consumers seen as passive victims of capitalism. Consumption is an active process and is mutually constituted with practices and processes of production, representation, identity and regulation. The themes and topics outlined in this book highlight the range and depth of analysis that has invigorated studies of cities and consumption. From structural global urban change to contextualised analysis of the micro-expressions of consumer practices in everyday life and the meanings that consumers invest in specific urban settings, studies of urban consumption are playing an important role in advancing understanding of both historical and contemporary urban life.

Consumption constitutes the 'stuff' that surrounds us all – advertising, television, entertainment, shopping – and it is consumption that underpins the images, sounds, smells and sights of the contemporary world. This argument can be contextualised by the work of theorists who suggest that ever-increasing and differentiated consumption opportunities have become the defining characteristic of twenty-first-century life. One of the main aims of the book has been to show that the study of cities and consumption allows a whole range of seemingly diverse but inter-connected elements to be brought together.

In a comprehensive review, Alan Warde (2002) seeks to summarise the advances made by studies of consumption. Warde describes ten ways in which consumption has been characterised, and in doing so offers a valuable template that not only summarises the debates that have been outlined in this book but also provides a guide for future research. First, then, Warde suggests that *consumption fosters meaningful work* in areas of non-necessity and that consumption can deliver bene-fits of non-alienating labour such as in hobbies and leisure. Second, *consumption promotes an aesthetic attitude*, in a world where generally people are in a better position to appreciate aesthetically appealing items and styles of living. This suggests that in everyday life necessity, survival and functional requirements do not determine the nature of people's surroundings. Moreover, although there remains a symbolic distinction between designer goods (which continue to act as a means of establishing social standing and status) and mass-produced goods, there is an increase in the importance of design in the latter. Warde also highlights the importance of mass media which make available things like film and music to people who exercise aesthetic judgements. Audiences of TV and advertising are reflective and critical about the content of production, and have an ability to keep up with fashions and constantly changing tastes.

Third, Warde shows that *consumption facilitates social rebellion* and expressing resistance in everyday life. Fashion, for example, has been used by young people to flout conventions and to express social dissatisfaction through clothing and other cultural manifestations. Fourth, *consumption is enjoyable and pleasurable* and hedonism is no longer totally morally frowned upon – pleasure and enjoyment have become pertinent human objectives. Warde suggests that the way in which consumption reliably provides pleasure should be applauded. Fifth, *consumption nurtures possessive individualism*, and commodified consumption rewards people through their private possessions and unconstrained use of goods. In these terms, consumption enhances self-esteem and feelings of security. However, Warde argues that possessive individualism has bred what some people believe to be a socially divisive competitive individualism, underpinned by conspicuous con-sumption and the pursuit of status, distinction and display. At the heart of this process is the private household, and the view that the rewards of the consumer culture can be attained through individual effort.

Sixth, *consumption supports socially meaningful practices*, and the purchasing of commodities is directly supporting socially meaningful practices such as gift exchange. Moreover, consumption can be implicitly or explicitly an expression of care, such as the provision of family meals. Hobbies and enthusiasm are a similar example that absorbs leisure time and can involve social participation that is beneficial to social relations in a civil society. Warde argues that participating in or attending a classic car rally, a professional football match or a sheepdog trial, and so on are means of confirming group membership and belonging. Seventh, *consumption supplies intellectual stimulation*, through television documentaries and films, newspapers, books, computer games, museums and theatrical performances. The psychological stimulation of new experiences such as travel and new technology is also seen as important.

Eighth, *consumption provides refreshing entertainment* and provides practical and mental diversions. Thus, the wasting of time is no longer associated with work not done or financial loss. Watching television is a great example, as a source of relaxation, of passing time, and of cessation of vigorous activity. Warde argues that consumption that fosters laziness and inactivity should be welcomed. Ninth, *consumption sustains comfort* through the raising of standards and comfort and cleanliness, and has reduced daily toil for most people. Mass consumption has indeed made life comfortable for those who are its beneficiaries, and a substantial majority in western societies are now unwilling to see those standards fall. Finally, *you can always get want you want* – consumer organisations are increasingly influential in ensuring that producers are subject to regulations that are centred on providing consumer satisfaction.

However, in his review of the different ways in which consumption has been depicted – as offering opportunities for self-development, independence, control, gratification, tolerance, comfort and as having a positive contribution to human condition – Alan Warde's review is also bound up with some reservations. Warde suggests, for example, that the role of individual choice in consumer behaviour is exaggerated. He suggests that the dominant view of consumption is that people buy what they personally want and thereby satisfy their individual material, symbolic and emotional needs and aspirations. However, Warde concludes that such a view is based on an understanding of consumption that presumes the sovereignty of consumer choice and reduces explanation to the reasons individuals make particular choices. Warde argues that behaviour is collective and situational, and if collective and institutional conditions of consumption are ignored then the structure of unequal distribution of power in the various fields of consumption is also overlooked. This viewpoint importantly stresses that not all people are attributed with an equal capacity for control over their situation.

Warde also highlights that much of consumption is about a dull compulsion, and that there is a mundane, unreflective ordinariness to consumption. This view is contra to the idea that consumption is a matter of decision and choice. Warde suggests that many consumption acts occur without mental input. He argues that we need to know more about the extent to which we follow a model of habit as opposed to conscious reflection. Moreover, Warde suggests that much of what we consume is largely impervious to status enhancement or fashion, and that processes such as habituation, routinisation and appropriation are key mechanisms in consumption. Finally, Warde argues that studies of consumption have predominantly fostered a complacent suspending of critique regarding consumer culture in general. Issues of social exclusion need to be addressed more fully. He argues that studies of consumption must seek to identify connections between social institutions implicated in sustaining contemporary consumption patterns (and their inequalities) and a more critical evaluation of the general social effects of the recent expansion of consumerism. Alan Warde's depiction of the ways in which consumption has been conceptualised, plus his concerns over theoretical and empirical shortcomings, provide a valuable review of the intellectual terrain. Warde's review provides a clear and concise case for the need to undertake focused and experiential research that links consumption to production, representation, identity and regulation.

It has been shown throughout this book that a wealth of diverse and interesting arguments and topics have stimulated the advances in our understanding of the relationship between cities and consumption. What has also been shown is that if such progress is to continue, future studies of consumption need to emulate this diversity, imagination and critical debate. In particular, research must engage with the ways in which consumption is an area of conflict; how and what we consume; and how everyday consumer practices are constituted. Such an agenda will augment the advances already made, and continue to articulate the diverse and complex ways in which consumption is central to the physical, political, economic, social, spatial and cultural practices and processes that affect the development of cities throughout the world.

This research agenda, however, not only offers a way of thinking about the relationship between cities and consumption by providing an understanding and examples of the way in which consumer culture positively and negatively impacts upon daily life, but also provides a way of critically assessing the nature of contemporary and future urban life. It has been argued throughout this book that it is not just commodities but also cities and the spaces and places within them that are laden with value and meaning. While global capitalism ensures that inequality and disadvantage are bound up in the relationship between consumption and urban life, the productive capabilities of consumption also offer opportunities for more progressive trends. A research agenda that captures both the enabling and the constraining nature of consumption is vital if we are to more fully understand the many ways in which consumption allows and disallows people and places from fully participating in the life of the city.

For example, throughout this book examples of division, inequality, conflict and marginalisation have been shown to be central to the relationship between cities and consumption. However, in contrast to such problematic political, economic, social and spatial disadvantage and segregation, positive examples of hybrid identities and socially inclusive consumption practices, new forms of urban sociability, and political activism generated through the interpenetration of consumption and daily life, have been depicted. For instance, popular concern regarding the westernisation of Singapore culture has led to an assertion of confidence in local consumer products and, moreover, legislation aimed at reducing personal debt due to excessive levels of credit, and hence the related socially divisive extremes of wealth and poverty. Discount stores have been shown to be places where consumption creates social mixing (the associated problems of low wages and debt economy notwithstanding). Farmers markets, car boot sales and thrift shops are examples of alternative consumption spaces that differentially offer more equitable relations between producer and consumer and also a sense of community, based around conviviality and personal relationships. Immigration, sexual identity, resilient ethnic consumption cultures, and young and older people marking out their place and visibly occupying the city through their consumption practices have been identified as useful examples of how people and social groups forge belonging, participation and identity through the relationship between consumption and urban life. The important (and often conflictual) relationship that consumption-led regeneration has with all these examples was also shown to be a key element of urban competitive advantage and how partnerships between private capital and central and local government impact on different people and places within the city.

It is clear that the relationship between consumption and urban life has contributed to social and spatial division within the city. However, more progressive exemplars of consumption practices representing (the pursuit or achievement of) more equitable political, economic, social and cultural practices can be found. While consumer culture is divisive, the productive nature of consumption, and the ways in which commodities and urban spaces and places can be ascribed value and meaning that can be interpreted, appropriated and re-appropriated in multiple different ways, offer the potential for more inclusive urban milieux. At present, consumption contributes more to underscoring structural inequalities and disadvantage, than to significantly ameliorating such divisions. The consumption-related practices and processes that have featured throughout this book, however, offer insights not only into the discursive and differential ways in which the relationship between inequality and urban life is constructed, but also into some of the ways in which this is being challenged and overcome.

> **Learning outcomes**
>
> - To understand and be able to describe the key themes covered in the book
> - To have a critical understanding of debates and case studies and be able to draw links across the chapters
> - To be aware of the future research agenda for studies of cities and consumption

## Further reading

David B. Clarke, Marcus Doel and Kate Housiaux (eds) (2003) *The Consumption Reader*, London: Routledge.

Danny Miller (ed.) (1995) *Acknowledging Consumption: A Review of New Studies*, London: Routledge.

Malcolm Miles and Tim Hall (eds) (2003) *Urban Futures: Critical Commentaries on Shaping the City*, London: Routledge.

Steven Miles, Alison Anderson and Kevin Meethan (eds) (2002) *The Changing Consumer: Markets and Meanings*, London: Routledge.

The contributions to these edited books represent a valuable collection of writing, from many different disciplines, and together provide a diverse and comprehensive range of topics relating to the study of cities and consumption.

# References

Adorno, T. and Horkheimer, T. W. (1973) *Dialectic of Enlightenment*, London: Allen Lane.

*The Advertiser* (1999a) 'Public art – you chose', (14/1/99), 1.

*The Advertiser* (1999b) 'Sounds of industry to pervade city', (13/5/99), 3.

*The Advertiser* (1999c) 'Bring touch of colour', (11/2/99), 1.

Allen, J., Massey, D. and Pryke, M. (eds) (1999) *Unsettling Cities*, London: Routledge.

Amin, A. (1994) *Post-Fordism: A Reader*, Oxford: Blackwell.

Amin, A. and Thrift, N. (2002) *Cities: Reimagining the Urban*, Cambridge: Polity.

Appadurai, A. (1996) *Modernity at Large*, Minneapolis, MN: University of Minnesota Press.

Arts Business Ltd (1993) *Major Touring Venues for Stoke-on-Trent*, London: Arts Business Ltd.

Arts Council of England (1995) *Arts Councils Study: Trends to 2006*, London: Arts Council of England.

Ashworth, G. J. and Voogod, H. (1994) 'Marketing and place promotion', in Gold, J. R. and Ward, S. V. (eds) *Place Promotion: The Use of Publicity and Marketing to Sell Towns and Regions*, Chichester: John Wiley, 39–52.

Back, L. (1996) *New Ethnicities and Urban Culture: Racisms and Multiculture in Young Lives*, London: University College London Press.

Badgett, M. V. L. (2001) *Money, Myths and Change: The Economic Lives of Lesbians and Gay Men*, Chicago, IL: University of Chicago Press.

Ball, W. and Beckford, J. A. (1997) 'Religion, education and city politics: a case study of community mobilisation', in Jewson, N. and McGregor, S. (eds) *Transforming Cities: Contested Governance and New Spatial Divisions*, London and New York: Routledge, 193–204.

Banfield, E. C. (1970) *The Unhealthy City: The Nature and Future of Our Urban Crisis*, London: Brown Little.

Barke, M. and Harrop, K. (1994) 'Selling the industrial town: identity, image and illusion', in Gold, J. R. and Ward, S. V. (eds) *Place Promotion: The Use of Publicity and Marketing to Sell Towns and Regions*, Chichester: John Wiley, 93–114.

Barthes, R. (1965) *Elements of Semiology*, New York: Hill and Wang.

Barthes, R. (1982) *Mythologies*, London: Cape.

Baudelaire, C. (1955) *Art in Paris 1845–1862: Salons and Other Exhibitions*, Paris: Phadlin.

Baudrillard, J. (1975) *The Mirror of Production*, St Louis, MO: Telos.

Baudrillard, J. (1981) *For a Critique of the Political Economy of the Sign*, St Louis, MO: Telos.

Baudrillard, J. (1988) *America*, London: Verso.

Baudrillard, J. (1993) *Symbolic Exchange and Death*, London: Sage.

Bauman, Z. (1987) *Legislators and Interpreters: On Modernity, Post-modernity and Intellectuals*, Cambridge: Polity Press.

Bauman, Z. (1998) *Work, Consumerism and the New Poor*, Buckingham: Open University Press.

Beazley, M., Loftman, P. and Nevin, B. (1997) 'Downtown redevelopment and community resistance: an international perspective', in Jewson, N. and MacGregor, S. (eds) *Transforming Cities: Contested Governance and New Spatial Divisions*, London: Routledge, 181–192.

Beck, U. (1992) *Risk Society: Towards a New Modernity*, London: Sage.

Bell, D. (1976) *The Coming of the Post-industrial Age*, London: Basic Books.

Bell, D. (2001) *Introduction to Cybercultures*, London: Routledge.

Bell, D. (2002) 'Fragments for a new urban culinary geography', *Journal for the Study of Food and Society* 6(1), 10–21.

Bell, D. (2005) 'Commensality, urbanity, hospitality', in Lashley, C., Lynch, P. and Morrison, A. (eds) *Critical Hospitality Studies*, Oxford: Butterworth Heinemann.

Bell, D. and Binnie, J. (2000) *The Sexual Citizen: Queer Politics and Beyond*, Cambridge: Polity Press.

Bell, D. and Binnie, J. (2003) 'Rethinking sexual citizenship in the city', paper presented at the annual meeting of the *Association of American Geographers*, New Orleans, LA, March.

Bell, D. and Valentine, G. (eds) (1995) *Mapping Desire: Geographies of Sexualities*, London: Routledge.

Bell, D. and Valentine, G. (1997) *Consuming Geographies: We Are Where We Eat*, London: Routledge.

Benjamin, W. (1973) *Illuminations*, London: Fontana.

Benjamin, W. (1999) *The Arcades Project*, trans. H. Eiland and K. McLaughlin, Cambridge, MA: Harvard.

Berman, M. (1992) *All That Is Solid Melts into Air: The Experience of Modernity*, London: Verso.

Bhabha, H. (1994) *The Location of Culture*, London: Routledge.

Bianchini, F. and Ghilardi, L. (2004) 'The culture of neighbourhoods: a European perspective', in Bell, D. and Jayne, M. (eds) *City of Quarters: Urban Villages in the Contemporary City*, Aldershot: Ashgate, 237–248.

Bianchini, F. and Parkinson, M. (eds) (1993) *Cultural Policy and Urban Regeneration: The Western European Experience*, Manchester: Manchester University Press.

Binnie, J. (1995) 'Trading places: consumption, sexuality and the production of queer space', in Bell, D. and Valentine, G. (eds) *Mapping Desire: Geographies of Sexualities*, London: Routledge, 182–199.

Binnie, J. (2000) 'Cosmopolitanism and the sexed city', in Bell, D. and Haddour, A. (eds) *City Visions*, Harlow: Prentice Hall, 166–178.

Binnie, J. (2004) 'Quartering sexualities: gay villages and sexual citizenship', in Bell, D. and Jayne, M. (eds) *City of Quarters: Urban Villages in the Contemporary City*, Aldershot: Ashgate, 163–172.

Binnie, J. and Skeggs, B. (2004) 'Cosmopolitan knowledge and the production and consumption of sexualised spaces: Manchester's gay village', *Sociological Review*, 18, 39–61.

Blackman, T. (1995) *Urban Policy in Practice*, London: Routledge.

Bocock, R. (1993) *Consumption*, London: Routledge.

Borden, I. (2001) *Skateboarding, Space and the City*, Oxford: Berg.

Bourdieu, P. (1984) *Distinction: A Social Critique of the Judgement of Taste*, London: Routledge and Kegan Paul.

Boyle, M. and Hughes, G. (1991) 'The politics of representation of the "real" discourses from the left in Glasgow's role of European City of Culture', *Area* 23, 3–28.

Brooker, P. (1999) *A Concise Glossary of Cultural Theory*, London: Arnold.

Brown, S. (1995) *Postmodern Marketing*, London and New York: Routledge.

Brownhill, S. (1994) 'Selling the inner city regeneration and place marketing in London's Docklands', in Gold, J. R. and Ward, S. V. (eds) *Place Promotion: The Use of Publicity and Marketing to Sell Towns and Regions*, Chichester: John Wiley, 132–151.

Buck-Morss, S. (1989) *The Dialectics of Seeing*, Cambridge, MA: MIT Press.

Campbell, C. (1989) *The Romantic Ethic and the Spirit of Modern Capitalism*, Oxford: Blackwell.

Campbell, C. (1995) 'The sociology of consumption', in Miller, D. (ed.) *Acknowledging Consumption: A Review of New Studies*, London: Routledge, 96–126.

Castells, M. (1977) *The Urban Question*, London: Edward Arnold.

Castells, M. (1978) *City, Class and Power*, London: Edward Arnold.

Castells, M. and Hall, P. (1994) *Technopoles of the World*, London: Routledge.

Chan, W. (2003) 'Finding Chinatown: ethnocentrism and urban planning', in Bell, D. and Jayne, M. (eds) *City of Quarters: Urban Villages in the Contemporary City*, Aldershot: Ashgate, 173–190.

Chang, T. C. (2000) 'Renaissance Revisited: Singapore as a "Global City for the Arts"', *International Journal of Urban and Regional Research,* 24(4), 818–831.

Chatterton, P. and Hollands, R. (2002) 'Theorising urban playscapes: producing, regulating and consuming youthful nightlife city spaces', *Urban Studies* 39(1), 95–116.

Chatterton, P. and Hollands, R. (2003) *Urban Nightscapes: Youth Culture, Pleasure Spaces and Corporate Power*, London: Routledge.

Chauncey, G. (1994) *Gay New York*, New York: Basic Books.

Chauncey, G. (1996) '"Privacy could only be had in public": gay uses of the streets', in Saunders, J. (ed.) *Stud: Architectures of Masculinity*, New York: Princeton Architectural Press, 244–267.

Chua, B-H. (1998) 'World cities, globalisation and the spread of consumerism: a view from Singapore', *Urban Studies*, 35(5–6), 981–1000.

City Centre Management (1998a) *Stoke-on-Trent City Centre: A Better, Safer Place to Be* (Video).

City Centre Management (1998b) *Action '98*.

Clarke, A. J. (1997) 'Tupperware: suburbia, sociality and mass consumption', in Silverstone, R. (ed.) *Visions of Suburbia*, London: Routledge, 132–160.

Clarke, D. B. (1997) 'Consumption and the city, modern and postmodern', *International Journal of Urban and Regional Research* 21(2), 218–237.

Clarke, D. B. (1998) 'Consumption, identity and space-time', *Consumption, Markets, Culture* 2(3), 233–258.

Clarke, D. B. (2003) *The Consumer Society and the Postmodern City*, London: Routledge.

Clarke, D. B. and Bradford, M. G. (1998) 'Public and private consumption and the city', *Urban Studies* 35(5–6), 856–888.

Clarke, D. B. and Purvis, M. (1994) 'Dialectics, difference and the geographies of consumption', *Environment and Planning A* 26, 1091–1109.

Clarke, D. B., Doel, M. and Housiaux, K. (eds) (2003) *The Consumption Reader*, London: Routledge.

Corrigan, P. (1997) *The Sociology of Consumption*, London: Sage.

Cox, K. (1995) 'Globalisation, competition and the politics of local economic development', *Urban Studies* (32), 213–225.

Crang, P. (1996) 'Displacement, consumption and identity', *Environment and Planning A* 28, 47–67.

Crang, P., Dwyer, C. and Jackson, P. (2004) 'Transnationalism and the spaces of commodity culture', *Progress in Human Geography* 27(4), 438–456.

Crewe, L. (2000) 'Geographies of retailing and consumption', *Progress in Human Geography* 24(2), 275–290.

Crewe, L. and Lowe, M. (1995) 'Gap on the map? Towards a geography of consumption and identity', *Environment and Planning A* 27, 1877–1898.

Crilley, D. (1993) 'Architecture and advertising: constructing the image of redevelopment', in Kearns, G. and Philo, C. (eds) *Selling Places: The City as Cultural Capital, Past and Present*, Oxford: Pergamon Press, 231–252.

Crompton, R. (1996) 'Consumption and class analysis', in Edgell, S., Heatherington, K. and Warde, A. (eds) *Consumption Matters: The Production and Experience of Consumption*, Oxford: Blackwell, 110–123.

*A Cultural Strategy for Stoke-on-Trent* (1991) London: Comedia Consultants.

Dant, T. and Martin, T. (2001) 'By car – carrying modern society', in Warde, A. and Gronow, M. (eds) *Ordinary Consumption*, London: Routledge, 143–175.

Davis, M. (1990) *City of Quartz: Excavating the Future in Los Angeles*, London and New York: Verso.

Davis, M. (1992) *City of Quartz*, New York: Vintage.

Dear, M. (2000) *The Postmodern Urban Condition*, Oxford: Blackwell.

de Certeau, M. (1984) *The Practices of Everyday Life*, Berkeley, CA: University of California Press.

de Certeau, M. (1988) *The Writing of History*, New York: Columbia University Press.

de Certeau, M. (1993) 'Walking in the city', in During, S. (ed.) *The Cultural Studies Reader*, London: Routledge, 207–224.

DeFilippis, J. (2004) 'Fables of the reconstruction (of the fables . . .): lower Manhattan after 9/11', in Bell, D. and Jayne, M. (eds) *City of Quarters: Urban Villages in the Contemporary City*, Aldershot: Ashgate, 56–70.

de Saussure, F. (1915) *Course in General Linguistics*, LaSalle, IL: Open Court Publishing Company.

Department of Culture, Media and Sport (2001a) *Culture and the Heart of Regeneration*, London: The Stationery Office.

Department of Culture, Media and Sport (2001b) *Time for Reform*, London: The Stationery Office.

Department of Environment, Transport and Planning (2000) *Towards an Urban Renaissance: The Report of the Urban Task Force*, London: The Stationery Office.

Domosh, M. (1996) 'The feminized retail landscape: gender ideology and consumer culture in nineteenth-century New York', in Wrigley, N. and Lowe, M. (eds) *Retailing, Capital and Consumption: Towards the New Retail Geography*, London: Longman, 257–270.

Donald, J. (1999) *Imaging the Modern City*, London: Athlone.

Douglas, V. and Shaw, D. (2001) 'The post-industrial city', in Paddison, R. (ed.) *Handbook of Urban Studies*, London: Sage, 284–295.

Du Gay, P. (ed.) (1997) *Production of Culture/Cultures of Production*, London: Sage.

Dwyer, C. and Crang, P. (2002) 'Fashioning ethnicities: the commercial spaces of multiculture', *Ethnicities* 2, 410–430.

Edensor, T. (1998) 'The culture of the Indian street', in Fyfe, N. R. (ed.) *Images of the Street: Planning, Identity and Control in Public Space*, London: Routledge, 205–224.

Edensor, T. (2003) 'M6 – Junction 16–19: defamiliarizing the mundane roadscape', *Space and Culture*, 6(2), 151–168.

Edwards, M. (1996) *Potters at Play*, Leek, Staffs: Churnet Valley Books.

Evans, G. (2001) *Cultural Planning: An Urban Renaissance?*, London: Routledge.

Featherstone, M. (1991) *Consumer Culture and Postmodernism*, London: Sage.

Finkelstein, J. (1998) 'Dining out: the hyperreality of appetite', in Scapp, R. and Seitz, J. (eds) *Eating Culture*, Albany, NY: SUNY Press, 201–215.

Fiske, J. (1989) *Reading the Popular*, London: Routledge.

Florida, R. (2002) *The Rise of the Creative Class*, New York: Basic Books.

Fokkema, T., Gierveld, J. and Nijkamp (1996) 'Big movies, big problems: reasons for the elderly to move?', *Urban Studies* (33)2, 353–377.

Foster, J. (1997) 'Challenging perceptions: "community" and neighbourliness on a difficult-to-let estate', in Jewson, N. and McGregor, S. (eds) *Transforming Cities: Contested Governance and New Spatial Divisions*, London and New York: Routledge, 56–69.

Frisby, D. (2001) *Cityscapes of Modernity*, Cambridge: Polity Press.

Fritzsche, P. (1996) *Reading Berlin*, Cambridge, MA: Harvard University Press.

Fyfe, N. R. (ed.) (1998) *Images of the Street: Planning, Identity and Control in Public Space*, London: Routledge.

Gardiner, M. E. (2000) *Critiques of Everyday Life*, London: Routledge.

Geddes, M. (1997) 'Poverty, excluded communities and local democracy', in Jewson, N. and McGregor, S. (eds) *Transforming Cities: Contested Governance and New Spatial Divisions*, London and New York: Routledge, 204–218.

Gibson, L. and Stevenson, L. (2004) 'Urban space and the uses of culture', *International Journal of Cultural Policy* 10(1), 1–4.

Giddens, A. (1973) 'Review of class struggle of advanced societies', *British Journal of Sociology* 29, 10–15.

Giddens, A. (1991) *The Consequences of Modernity*, Cambridge: Polity Press.

Gilroy, P. (1987) *There Ain't No Black in the Union Jack*, London: Unwin Hyman.

Gilroy, P. (1993) *Small Acts: Thoughts on the Politics of Black Cultures*, London and New York: Serpent's Tale.

Glennie, P. D. (1995) 'Consumption within historical studies', in Miller, D. (ed.) *Acknowledging Consumption: A Review of New Studies*, London: Routledge, 164–203.

Glennie, P. D. (1998) 'Consumption, consumerism and urban form: historical perspectives', *Urban Studies* 35(5–6), 927–951.

Glennie, P. D. and Thrift, N. J. (1992) 'Modernity, urbanism and modern consumption', *Environment and Planning A* 10, 432–443.

Glennie, P. D. and Thrift, N. J. (1996) 'Consumers, identities and consumption spaces in early modern England', *Environment and Planning A* 28, 25–45.

Gluckman, A. and Reed, B. (eds) (1997) *Homo Economics: Capitalism, Community and Lesbian and Gay Life*, New York: Routledge.

Gold, J. R. (1994) 'Locating the message: place promotion as image and communication', in Gold, J. R. and Ward, S. V. (eds) *Place Promotion: The Use of Publicity and Marketing to Sell Towns and Regions*, Chichester: John Wiley, 19–37.

Gold, J. R. and Ward, S. V. (eds) (1994) *Place Promotion: The Use of Publicity and Marketing to Sell Towns and Regions*, Chichester: John Wiley.

Goldthorpe, J. H., Lockwood, D., Bechhofer, F. and Platt, J. (1969) *The Affluent Worker: Industrial Attitudes and Behaviour*, Cambridge: Cambridge University Press.

Goody, B. (1994) 'Art-full places: public art to sell public spaces', in Gold, J. R. and Ward, S. V. (eds) *Place Promotion: The Use of Publicity and Marketing to Sell Towns and Regions*, Chichester: John Wiley, 151–179.

Gottdiener, M. (ed.) (2000) *New Forms of Consumption: Consumers, Culture and Commodification*, New York: Rowman and Littlefield.

Gottdiener, M. and Lagopoulos, A. (eds) (1986) *The City and the Sign: An Introduction to Urban Semiotics*, Guilford, NY: Columbia University Press.

Gregory, D. and Walford, R. (eds) (1988) *Horizons in Human Geography*, London: Rowman and Littlefield.

Gregson, N. and Crewe, L. (2003) *Second-Hand Cultures*, Oxford: Berg.

Gronow, J. and Warde, A. (eds) (2001) *Ordinary Consumption*, London: Routledge.

Guy, C. M. (1996) 'Corporate strategies in food retailing and their local impacts', *Environment and Planning A* 28, 1575–1602.

Hage, G. (1997) 'At home in the entrails of the west', in Grace, H., Hage, G., Johnson, L., Langsworth, J. and Symonds, M. *Home/World: Space, Community and Marginality in Sydney's West*, Annandale, NSW: Pluto Press, 99–153.

Hall, P. (1995) 'Urban stress, creative tension', *The Independent* (21/2/95), 15.

Hall, P. (2000) 'Creative cities and economic development', *Urban Studies* 37(4), 639–649.

Hall, S. (1973) 'Encoding/decoding', in Centre for Contemporary Cultural Studies (ed.) *Culture, Media, Language: Working Papers in Cultural Studies, 1972–79*, London: Hutchinson, 128–38.

Hall, T. (1995) 'The second industrial revolution: cultural reconstruction of industrial regions', *Landscape Research* 20, 112–123.

Hall, T. (1997) '(Re)placing the city: cultural relocation and the city centre', in Westwood, S. and Williams, J. (eds) *Imagining Cities: Scripts, Signs, Memories*, London: Routledge, 202–218.

Hall, T. (1998) *Urban Geography*, London: Routledge.

Hall, T. (2003) 'Car-ceral cities: social geographies of everyday urban mobility', in Miles, M. and Hall, T. (eds) *Urban Futures: Critical Commentaries on Shaping the City*, London: Routledge, 92–107.

Hall, T. and Hubbard, P. (eds) (1998) *The Entrepreneurial City: Geographies of Politics, Regimes and Representation,* London: John Wiley.

Hambleton, R. (1996) 'Future directions for urban government in Britain and America', in LeGates, R. T. and Stout, F. (eds) *The City Reader*, London: Routledge, 282–292.

Handy, C. (1994) *The Empty Raincoat*, London: Arrow Business.

Harvey, D. (1973) *Social Justice and the City*, Baltimore, MD: Johns Hopkins University.

Harvey, D. (1985) 'The geopolitics of capitalism', in Gregory, D. and Urry, J. (eds) *Social Relations and Spatial Structures*, London: Macmillan, 46–57.

Harvey, D. (1989a) *The Condition of Postmodernity*, Baltimore, MD: Johns Hopkins University.

Harvey, D. (1989b) *The Urban Experience*, Baltimore, MD: Johns Hopkins University.

Harvey, D. (1997) 'Contested cities: social processes and spatial form', in Jewson, N. and McGregor, S. (eds) *Transforming Cities: Contested Governance and New Spatial Divisions*, London: Routledge, 19–27.

Hebdige, D. (1979) *Subculture: The Meaning of Style*, London: Methuen.

Hennessy, R. (2000) *Profit and Pleasure: Sexual Identities in Late Capitalism*, London: Taylor and Francis.

Herrschel, T. (1998) 'From socialism to post-Fordism: the local state and economic politics in East Germany', in Hall, T. and Hubbard, P. (eds) *The Entrepreneurial City: Geographies of Politics, Regime and Representation*, Chichester: John Wiley, 173–196.

Hewison, R. (1995) *The Culture of Consensus*, London: Methuen.

Holcomb, B. (1994) 'City make-overs: marketing the post-industrial city', in Gold, J. R. and Ward, S. V. (eds) *Place Promotion: The Use of Publicity and Marketing to Sell Towns and Regions*, Chichester: John Wiley, 115–131.

Holliday, R. (2005) 'Home truths?', in Bell, D. and Hollows, J. (eds) *Ordinary Lifestyles: Popular Media, Consumption and Taste*, Milton Keynes: Open University Press, 27–34.

Holliday, R. and Jayne, M. (2000) 'The potters holiday', in Edensor, T. (ed.) *Reclaiming Stoke-on-Trent: Leisure, Space and Identity in The Potteries*, Stoke-on-Trent: Staffordshire University Press, 117–200.

Hubbard, P. (1998) 'Introduction: representation, culture and identities', in Hall, T. and Hubbard, P. (eds) *The Entrepreneurial City: Geographies of Politics, Regime and Representation*, Chichester: John Wiley, 198–201.

Hubbard, P. (2002) 'Screen-shifting: consumption, "riskless risks" and the changing geographies of cinema', *Environment and Planning A* 34, 1239–1258.

Hughes, G. (1997) 'Policing late modernity: changing strategies of crime management in contemporary Britain', in Jewson, N. and McGregor, S. (eds) *Transforming Cities: Contested Governance and New Spatial Divisions*, London: Routledge, 153–165.

Ilmonen, J. (2001) 'Routinization or reflexivity? Consumers and normative claims for environmental consideration', in Gronow, J. and Warde, A. (eds) *Ordinary Consumption*, London: Routledge, 24–36.

Jackson, P. (1997) 'Domesticating the street: the contested spaces of the High Street', in Fyfe, N. (ed.) *Images of the Street: Planning, Identity and Control of Public Space*, London: Routledge, 176–191.

Jackson, P. (1998) 'Constructions of "whiteness" in the geographical imagination', *Area* 30(2), 99–106.

Jackson, P. (2000) 'Guest Editorial', *Environment and Planning A* 27, 1875–1876.

Jackson, P. and Holdbrook, B. (1995) 'Multiple meanings, shopping and the cultural politics of identity', *Environment and Planning A* 27, 1913–1930.

Jackson, P. and Thrift, N. (1995) 'Geographies of consumption', in Miller, D. (ed.) *Acknowledging Consumption: A Review of New Studies*, London: Routledge, 204–237.

Jacobs, B. (1992) *Fractured Cities: Capitalism, Community and Empowerment in Britain and America*, London and New York: Routledge.

Jameson, F. (1991) *Postmodernism or the Cultural Logic of Late Capitalism*, Durham, NY: Duke University Press.

Jansson, R. (2003) 'The negotiated city image: symbolic representation and change through urban consumption', *Urban Studies* 40(3), 463–479.

Jarvis, B. (1994) 'Transitory topographies: place, events, promotions and propaganda', in Gold, J. R. and Ward, S. V. (eds) *Place Promotion: The Use of Publicity and Marketing to Sell Towns and Regions*, Chichester: John Wiley, 181–193.

Jayne, M. (2000) 'Imag(in)ing a post-industrial potteries', in Bell, D. and Haddour, A. (eds) *City Visions*, Harlow: Prentice Hall, 12–26.

Jayne, M. (2003) 'Too many voices, "too problematic to be plausible": representing multiple responses to local economic development strategies', *Environment and Planning A* 35, 959–981.

Jayne, M. (2005) 'Creative industries: the regional dimension', *Environment and Planning C: Government and Policy* 23, 537–556.

Jessop, B. (1997) 'The entrepreneurial city: re-imagining localities, re-designing economic government, or re-structuring capital?', in Jewson, N. and McGregor, D. (eds) *Transforming Cities: Contested Governance and New Spatial Divisions*, London: Routledge, 19–27.

Jessop, B. (1998) 'The narrative of enterprise and the enterprise of narrative: place marketing and the entrepreneurial city', in Hall, T. and Hubbard, P. (eds) *The*

*Entrepreneurial City: Geographies of Politics, Regime and Representation*, Chichester: John Wiley, 75–99.

Keith, M. and Pile, S. (eds) (1993) *Place and the Politics of Identity*, London: Routledge.

King, A. (ed.) (1996) *Re-presenting the City: Ethnicity, Capital and Culture in the 21st Century Metropolis*, Basingstoke: Macmillan.

Knopp, L. (1995) 'Sexuality and urban space: a framework for analysis', in Bell, D. and Valentine, G. (eds) *Mapping Desire: Geographies of Sexualities*, London: Routledge, 149–163.

Knox, P. (1987) 'The social production of the built environment: architects, architecture and the postmodern city', *Progress in Human Geography* 21(3), 154–377.

Kracauer, S. (1926) *The Mass Ornament*, Cambridge, MA: Harvard University Press.

Kumar, K. (1995) *From Post-Industrial to Post-Modern Society*, Oxford: Blackwell.

Lally, E. (2002) *At Home with Computers*, Oxford: Berg.

Landry, C. (1995) *The Creative City*, London: Demos Comedia.

Lash, S. (1990) *Sociology of Postmodernism*, London: Routledge.

Lash, S. (1999) *Another Modernity: A Different Rationality*, Oxford: Blackwell.

Lash, S. and Urry, J. (1987) *The End of Organised Capitalism*, Cambridge: Polity Press.

Lash, S. and Urry, J. (1994) *Economies of Sign and Space*, London: Sage.

Latham, A. (2003) 'Urbanity, lifestyle and making sense of the new urban cultural economy: notes from Auckland, New Zealand', *Urban Studies* 40(9), 1699–1724.

Laurier, E. and Philo, P. (2004) *The Cappuccino Community: Cafes and Civic Life in the Contemporary City*, (draft) published by the Department of Geography and Topographical Science, University of Glasgow at http://www.geog.gla.ac.uk/olpapers/elaurier002.pdf.

Lea, J. (1997) 'Post-Fordism and criminality', in Jewson, N. and McGregor, D. (eds) *Transforming Cities: Contested Governance and New Spatial Divisions*, London: Routledge, 42–55.

Leach, R. W. (1984) 'Transformation in a culture of consumption: women and department stores 1890–1925', *Journal of American History* 77, 319–342.

Lefebvre, H. (1971) *Everyday Life in the Modern World*, London: Allen Lane.

Lefebvre, H. (1984) *Everyday Life in the Modern World*, London: Transaction Publishers.

Lefebvre, H. (1991) *The Production of Space*, Oxford: Blackwell.

LeGates, R. T. and Stout, F. (eds) (1996) *The City Reader*, London: Routledge.

Leitner, H. (1990) 'Cities in pursuit of economic growth: the local state as entrepreneur', *Political Geography Quarterly* 9, 146–170.

Leitner, H. and Sheppard, E. (1998) 'Economic uncertainty, inter-urban competition and the efficacy of entreprenurialism', in Hall, T. and Hubbard, P. (eds) *The Entrepreneurial City: Geographies of Politics, Regime and Representation*, Chichester: John Wiley, 285–306.

Lever, W. F. (2001) 'The post-Fordist city', in Paddison, R. (ed.) *Handbook of Urban Studies*, London: Sage, 272–283.

Ley, D. and Olds, K. (1988) 'Landscape as spectacle: world's fairs and the culture of heroic consumption', *Environment and Planning D: Society and Space* 6(2), 191–212.

Loftman, P. and Nevin, B. (1998) 'Pro-growth local economic strategies: civic promotion and local needs in Britain's second city', in Hall, T. and Hubbard, P. (eds) *The Entrepreneurial City: Geographies of Politics, Regime and Representation*, Chichester: John Wiley, 129–148.

Logan, J. and Molotch, H. (1998) *Urban Fortunes: The Political Economy of Place*, Berkeley, CA: University of California Press.

Lury, C. (1996) *Consumer Culture*, Cambridge: Polity Press.

Mackay, D. (1991) 'Urban design and cultural interface', *Urban Design Quarterly: Where Urban Design Meets Culture* 37, 5–10.

Mackay, H. (ed.) (1997) *Consumption and Everyday Life*, London: Sage.

McKendrick, N., Brewer, J. and Plumb, J. H. (1982) *The Birth of a Consumer Society: The Commercialisation of Eighteenth Century England*, London: Europa.

Maenpaa, P. (2001) 'Mobile communication as a way of urban life', in Warde, A. and Gronow, J. (eds) *Ordinary Consumption*, London: Routledge, 49–60.

Maffesoli, M. (1992) *The Time of Tribes: The Decline of Individualism in Mass Society*, London: Sage.

Mahony, E. (1997) 'The people in parentheses: spaces under pressure in the post-modern city', in Clarke, D. B. (eds) *The Cinematic City*, London: Routledge, 168–185.

Malcolmson, R. (1973) *Popular Recreations in English Society 1700–1850*, Cambridge: Cambridge University Press.

Marsden, T. and Wrigley, N. (1995) 'Regulation, retailing and consumption', *Environment and Planning A* 27, 1899–1912.

Massey, D. (1993) 'Power-geometry and a progressive sense of place', in Bird, J., Curtis, B., Putnam, T., Robertson, G. and Tickner, L. (eds) *Mapping the Futures: Local Cultures, Global Change*, London and New York: Routledge, 54–69.

Massey, D. (1999) 'Cities in the world', in Massey, D., Allan, J. and Pile, S. (eds) *City Worlds*, London: Routledge, 32–46.

Mellor, R. (1997) 'Cool times for a changing city', in Jewson, N. and McGregor, D. (eds) *Transforming Cities: Contested Governance and New Spatial Divisions*, London: Routledge, 56–69.

Mercer, C. (2000) 'Cities and the wealth of nations: national policies and regional realities', conference paper presented at 'Cultural Industries and the City', Manchester Institute of Popular Culture, Manchester Metropolitan University.

Merrifield, A. (2000) 'The dialectics of dystopia: disorder and zero tolerance in the city', *International Journal of Urban Regional Research* 26(2), 473–489.

Miles, M. (1997) *Art, Space and the City: Public Art and Urban Futures*, London: Routledge.

Miles, M. (2004) 'Drawn and quartered: El Ravel and the Haussmannization of Barcelona', in Bell, D. and Jayne, M. (eds) *City of Quarters: Urban Villages in the Contemporary City*, Aldershot: Ashgate, 37–55.

Miles, M. and Hall, T. (eds) (2003) *Urban Futures: Critical Commentaries on Shaping the City*, London: Routledge.

Miles, S. (1998a) 'The consuming paradox: a new research agenda for urban consumption', *Urban Studies* 35(5–6), 1001–1008.

Miles, S. (1998b) *Consumerism as a Way of Life*, London: Sage.

Miles, S. (2001) *Social Theory in the Real World*, London: Sage.

Miles, S. and Paddison, R. (1998) 'Urban consumption: an historical note', *Urban Studies* 35(5–6), 815–832.

Miles, S., Anderson, A. and Meethan, K. (eds) (2002) *The Changing Consumer: Markets and Meanings*, London: Routledge.

Milestone, K. (1996) 'Regional variations: northernness and new urban economies of hedonism', in O'Connor, J. and Wynne, D. (eds) *From the Margins to the Centre: Cultural Production and Consumption in the Post-industrial City*, Aldershot: Arena, 91–115.

Miller, D. (ed.) (1995) *Acknowledging Consumption: A Review of New Studies*, London: Routledge.

Miller, D. (1998) *A Theory of Shopping*, Cambridge: Polity Press.

Mollencopft, J. (1983) *The Contested City*, Princeton, NJ: Princeton University Press.

Molotch, H. (1996) 'The political economy of growth machines', *Journal of Urban Affairs* 15(1), 29–53.

Molotch, H. (2003) *Where Stuff Comes From: How Toasters, Toilets, Cars, Computers and Many Other Things Come to Be as They Are*, London: Routledge.

Mooney, J. (1997) 'Violence, space and gender: the social and spatial parameters of violence against women and men', in Jewson, N. and McGregor, D. (eds) *Transforming Cities: Contested Governance and New Spatial Divisions*, London: Routledge, 100–115.

Moore, R. (2000) 'Poverty and partnership in the third European poverty programme', in Jewson, N. and McGregor, D. (eds) *Transforming Cities: Contested Governance and New Spatial Divisions*, London: Routledge, 166–178.

Mort, F. (1996) *Cultures of Consumption*, London: Routledge.

Mort, F. (1998) 'Consumption, masculinities and the mapping of London since 1950', *Urban Studies* 35(5–6), 889–907.

Munt, S. (1995) 'The lesbian flâneur', in Bell, D. and Valentine, V. (eds) *Mapping Desire: Geographies of Sexualities*, London: Routledge, 355–372.

Myers-Jones, H. J. and Brooker-Gross, S. R. (1994) 'Newspapers as promotional strategies for regional definition', in Gold, J. R. and Ward, S. V. (eds) *Place Promotion: The Use of Publicity and Marketing to Sell Towns and Regions*, Chichester: John Wiley, 193–212.

Nixon, S. (1996) *Hard Looks: Masculinities, Spectatorship and Contemporary Consumption*, London: Palgrave.

Noon, D., Smith-Canham, J. and England, M. (2000) 'Economic regeneration and funding', in Roberts, P. and Sykes, H. (eds) *Urban Regeneration: A Handbook*, London: Sage, 61–85.

Oatley, N. (ed.) (1998) *Cities, Economic Competition and Urban Policy*, London: Paul Chapman Publishing.

O'Connor, J. (1998) 'Popular culture, cultural intermediaries and urban regeneration', in Hall, T. and Hubbard, P. (eds) *The Entrepreneurial City: Geographies of Politics, Regime and Representation*, Chichester: John Wiley, 225–239.

O'Connor, J. and Wynne, D. (eds) (1996) *From the Margins to the Centre: Cultural Production and Consumption in the Post-industrial City*, Aldershot: Arena.

Pacione, M. (1997) *Britain's Cities: Geographies of Divisions in Urban Britain*, London and New York: Routledge.

Painter, J. (1997) 'Entrepreneurs are made not born: learning and urban regimes in the production of entrepreneurial cities', in Hall, T. and Hubbard, P. (eds) *The Entrepreneurial City: Geographies of Politics, Regime and Representation*, Chichester: John Wiley, 259–273.

Parker, D. (2000) 'The Chinese takeaway and the diasporic habitus: space, time and power geometries', in Hesse, B. (ed.) *Unsettled Multiculturalisms: Diasporas, Entanglements and Transruptions*, New York: Zed Books.

Parker, M. (1996) 'Shopping for principles: writing about Stoke-on-Trent's Festival Park', *Transgressions* 2/3, 38–54.

Philo, C. and Kearns, G. (1993) *Selling Places: City as Cultural Capital, Past and Present*, London: Architectural Press.

Pile, S. (1999) 'What is a city?', in Massey, D., Allan, J. and Pile, S. (eds) *City Worlds*, London: Routledge.

Pile, S., Brook, C. and Mooney, G. (eds) (1999) *Unruly Cities? Order/Disorder*, London: Routledge.

Pred, A. (1996) 'Interfusions: consumption, identity and the practices and power relations in everyday life', *Environment and Planning A* 28, 11–24.

Quilley, S. (1999) 'Entrepreneurial Manchester: the genesis of elite consciousness', *Antipode* 31(2), 185–211.

Radcliffe, P. (1997) '"Race", housing and the city', in Jewson, N. and McGregor, D. (eds) *Transforming Cities: Contested Governance and New Spatial Divisions*, London: Routledge, 56–69.

Randall, S. (1995) 'City pride – from "municipal socialism" to "municipal capitalism"?', *Critical Social Policy* 15, 40–59.

Reekie, G. (1992) 'Changes in the Adamless Eden: the spatial and sexual transformation of a Brisbane department store 1930–90', in Shields, R. (ed.) *Lifestyle Shopping: The Subject of Consumption*, London: Routledge, 170–194.

Rendell, J. (2002) *Pursuit of Pleasure: Gender, Space and Architecture in Regency London*, London: Continuum.

Rendell, J., Penner, B. and Borden, I. (eds) (1999) *Gender, Space, Architecture: An Interdisciplinary Introduction*, London: Routledge.

Revill, G. (1994) 'Promoting the Forest of Dean: art ecology and the industrial landscape', in Gold, J. R. and Ward, S. V. (eds) *Place Promotion: The Use of Publicity and Marketing to Sell Towns and Regions*, Chichester: John Wiley, 232–245.

Ritzer, G. (1998) *The McDonaldisation Thesis*, London: Sage.

Ross, K. (1996) *Fast Cars, Clean Bodies: Decolonization and the Reordering of French Culture*, Cambridge, MA: MIT Press.

Rothman, H. (2002) *Neon Metropolis: How Las Vegas Started the Twenty-First Century*, New York: Taylor and Francis.

Ryan, J. and Fitzpatrick, H. (1996) 'The spaces that difference makes: negotiation and urban identities through consumption practices', in O'Connor, J. and Wynne, D. (eds) *From the Margins to the Centre: Cultural Production and Consumption in the Post-industrial City*, Aldershot: Arena, 169–201.

Sassatelli, S. (2001) 'Smart Life 9.0: representations of everyday life in future studies', in Gronow, J. and Warde, A. (eds) *Ordinary Consumption*, London: Routledge, 36–49.

Saunders, P. (1981) *Social Theory and the Urban Question*, London: Hutchinson.

Savage, M. and Warde, A. (1993) *Sociology, Capitalism and Modernity*, New York: Continuum.

Scott, A.J. (2000) *The Cultural Economy of Cities: Essays on the Geographies of Image Producing Industries*, London: Sage.

Sennett, R. (1977) *The Fall of Public Man*, Cambridge: Cambridge University Press.

Sennett, R. (1995) *Flesh and Stone*, London: Faber and Faber.

*The Sentinel* (1999) 'Stoke's Relocation Rating', (14/9/99), 26.

Sheaff, M. (1997) 'Urban partnerships, economic regeneration and the healthy city', in Jewson, N. and McGregor, D. (eds) *Transforming Cities: Contested Governance and New Spatial Divisions*, London: Routledge, 141–152.

Sheller, M. and Urry, J. (2000) 'The city and the car', *International Journal of Urban and Regional Research* 27, 737–757.

Shields, R. (1991) *Places on the Margin*, London: Routledge.

Shields, R. (ed.) (1992a) *Lifestyle Shopping: The Subject of Consumption*, London: Routledge.

Shields, R. (1992b) 'Modernity, urban consumption and consumerism', *Environment and Planning D: Society and Space* 11, 599–601.

Shields, R. (1992c) 'A truant proximity: presence and absence in the space of modernity', *Environment and Planning D: Society and Space* 10, 181–198.

Shields, R. (1994) 'Fancy footwear: Walter Benjamin's notes on flâneurs', in Tister, K. (ed.) *The Flâneur*, London and New York: Routledge, 61–80.

Short, J. R. (1996) *The Urban Order: An Introduction to Cities, Culture and Power*, Oxford: Blackwell.

Short, J. R. and Kim, Y-H. (1998) 'Urban crisis/urban regeneration: selling the city in difficult times', in Hall, T. and Hubbard, P. (eds) *The Entrepreneurial City: Geographies of Politics, Regime and Representation*, Chichester: John Wiley, 55–75.

Simmel, G. (1907) *The Philosophy of Money* (1999, 2nd edn), London: Routledge.

Simmel, G. (1957) *Metropolis and Mental Life*, London: Blackwell.

Skeggs, B. (1997) *Formations of Class and Gender: Becoming Respectable*, London: Sage.

Slater, D. (1997) *Consumer Culture and Modernity*, Cambridge: Polity Press.

Smyth, H. (1994) *Marketing the City: The Role of Flagship Developments in Urban Regeneration*, London: E and FN Spon.

Soja, E. (1989) *Postmodern Geographies: The Reassertion of Space in Critical Social Theory*, London: Verso.

Soja, E. (1996) *Thirdspace: Journey to Los Angeles and Other Real-and-Imagined Places*, Oxford: Blackwell.

Southerton, D. (2001) 'Lifestyle and social integration: a study of middle class culture in

Manchester', in Gronow, J. and Warde, A. (eds) *Ordinary Consumption*, London: Routledge, 87–99.

Stallybrass, P. and White, A. (1986) *The Politics and Poetics of Transgression*, London: Methuen.

Stevenson, D. (2003) *Cities and Urban Cultures*, London: Sage.

Stone, C. (1989) *Regime Politics: Governing Atlanta 1946–1988*, Lawrence, KS: University of Kansas Press.

Taylor, F. (1911) *The Principles of Scientific Management*, Mineloa, NY: Dover Publications.

Taylor, I., Evans, I. and Fraser, P. (1996) *A Tale of Two Cities: Global Change, Local Feeling and Everyday Life in the North of England: A Study of Sheffield and Manchester*, London: Routledge.

Thorpe, M. (1999) 'Marginalisation and resistance through the prism of retirement', in Hearn, J. and Roseneil, S. (eds) *Consuming Cultures: Power and Resistance*, Basingstoke: Macmillan, 109–130.

Thrift, N. (2000) 'Not a straight line but a curve, or, cities are not mirrors of modernity', in Bell, D. and Haddour, A. (eds) *City Visions*, Harlow: Pearson Education, 233–264.

Tister, K. (ed.) (1994) *The Flâneur*, London: Routledge.

Tomlinson, A. (1990) *Consumption, Identity and Style: Marketing, Meaning and the Packaging of Pleasure*, London: Routledge.

Urry, J. (1990) *The Tourist Gaze*, London: Sage.

Urry, J. (1995) *Consuming Places*, London: Routledge.

Usborne, D. (2004) 'They think big in the Windy City. Here's the art to prove it', *The Independent*, (12/3/04), 17.

Valentine, G. (1989) 'The geography of women's fear', *Area* 21(4), 385–390.

Valentine, G. (1998) ' Food and the production of the civilised street', in Fyfe, N. (ed.) *Images of the Street: Planning, Identity and Control of Public Space*, London: Routledge, 192–204.

Veblen, T. (1899) *The Theory of the Leisure Class* (1994, 2nd edn), London: Constable.

Ward, S. V. (1994) 'Time and place: key themes in place promotion in USA, Canada and Britain since 1870', in Gold, J. R. and Ward, S. V. (eds) *Place Promotion: The Use of Publicity and Marketing to Sell Towns and Regions*, Chichester: John Wiley, 53–74.

Ward, S. V. (1998) 'Place marketing: a historical comparison of Britain and North America', in Hall, T. and Hubbard, P. (eds*)* *The Entrepreneurial City: Geographies of Politics, Regime and Representation*, Chichester: John Wiley, 29–53.

Warde, A. (1997) *Consumption, Food and Taste*, London: Sage.

Warde, A. (2002) 'Setting the scene: changing conceptions of consumption', in Miles, S., Anderson, A. and Meethan, K. (eds) *The Changing Consumer: Markets and Meanings*, London: Routledge, 10–24.

Williams, R. (1982) *Dream Worlds*, Berkeley, CA: University of California Press.

Wilson, E. (1992) *The Sphinx in the City*, Los Angeles, CA: University of California Press.

Wolff, T. (1985) 'The invisible flâneuse: women in the literature of modernity', *Theory, Culture and Society* 2, 27–40.

Woodward, I. (2003) 'Divergent narratives in the imagining of the home amongst middle-class consumers: aesthetics, comfort and symbolic boundaries of the self and home', *Journal of Sociology* 39(4), 391–412.

Wrigley, N. and Lowe, M. (1996) *Retailing, Consumption and Capital*, Harlow: Longman.

Wynne, D. and O'Connor, J. (1998) 'Consumption and the postmodern city', *Urban Studies* 35, 841–864.

Young, C. and Lever, J. (1997) 'Place promotion, economic location and the consumption of city image', *Tijdschrift voor Economische en Sociale Geographie* 88(4), 332–341.

Zukin, S. (1982) *Loft Living: Culture and Capital in Urban Change*, Baltimore, MD: Johns Hopkins University Press.

Zukin, S. (1991) *Landscapes of Power: From Detroit to Disney World*, Berkeley and Los Angeles, CA: University of California Press.

Zukin, S. (1995) *The Cultures of Cities*, Oxford and Cambridge, MA: Blackwell.

Zukin, S. (1998a) 'Urban lifestyles: diversity and standardisation in spaces of consumption', *Urban Studies* 35, 825–839.

Zukin, S. (1998b) 'From Coney Island to Las Vegas in the urban imaginary: discursive practices of growth and decline', *Urban Affairs Review* 33, 625–653.

Zukin, S. (2000) 'Space and symbols in an age of decline', in Miles, M., Hall, T. and Borden, I. (eds) *The City Cultures Reader*, London: Routledge, 81–94.

Zukin, S. (2004) *Point of Purchase: How Shopping Changed American Culture*, London: Routledge.

# Index